*LINEAR CIRCUITS
FOR
ELECTRONICS
TECHNOLOGY*

PRENTICE-HALL SERIES IN ELECTRONIC TECHNOLOGY

Dr. Irving L. Kosow, *editor*

Charles M. Thomson and Joseph J. Gershon, *consulting editors*

LINEAR CIRCUITS FOR ELECTRONICS TECHNOLOGY

GARY M. MILLER
Monroe Community College
Rochester, New York

PRENTICE-HALL, Inc.   Englewood Cliffs, New Jersey

Library of Congress Cataloging in Publication Data

MILLER, GARY M.
  Linear circuits for electronics technology.

  1. Electronic circuits.   I. Title.
TK7867.M46   621.3815'3   73-9632
ISBN 0-13-536698-4

© 1974 by
PRENTICE-HALL, Inc.
Englewood Cliffs, N.J.

All rights reserved. No part of this book may be reproduced in any form or by any means without permission in writing from the publisher.

10 9 8 7 6 5 4 3 2 1

Printed in the United States of America

PRENTICE-HALL INTERNATIONAL, INC., *London*
PRENTICE-HALL OF AUSTRALIA, PTY. LTD., *Sydney*
PRENTICE-HALL OF CANADA, LTD., *Toronto*
PRENTICE-HALL OF INDIA PRIVATE LIMITED, *New Delhi*
PRENTICE-HALL OF JAPAN, INC., *Tokyo*

To Jean, Mom, Dad, and Linda

Contents

Preface xv

1 Introductory Concepts 1

1-1 Introduction 1
1-2 P-N Junction Diode 2
1-3 Transistor Action 6
1-4 Amplification 13
1-5 Basic Transistor Amplification 15
1-6 Gain-Impedance Relations 17
   Problems 19

2 Rectification and Filters 23

2-1 Introduction 23
2-2 Half-Wave Rectification 24
2-3 Full-Wave Rectification 28
2-4 Filters for Power Supplies 31
2-5 Practical Filter Design 34
2-6 Voltage Multipliers 38
   Problems 40

3 Electronic Regulators 41

- 3–1 Introduction 41
- 3–2 Ideal Voltage Source 42
- 3–3 Zener Diode Shunt Regulators 46
- 3–4 Series Regulators 49
- 3–5 Variable Output Regulated Power Supply with Over-current Protection 53
- 3–6 Constant Current Supplies 57
  Problems 60

4 Basic Transistor Amplifiers 63

- 4–1 Introduction 63
- 4–2 Common-Base Amplifiers—DC Relations 63
- 4–3 Equivalent Circuit for the CB Transistor Amplifier 65
- 4–4 Common-Base Amplifiers 67
- 4–5 Common-Collector Configuration 72
- 4–6 Common-Collector Applications 74
- 4–7 Common-Emitter Amplifiers—Introduction 79
- 4–8 Common-Emitter Calculations 80
  Problems 85

5 Single-Stage Feedback Amplifiers 87

- 5–1 Introduction 87
- 5–2 Feedback Principles 88
- 5–3 Emitter Feedback Resistor 91
- 5–4 Gain—Impedance Relations 94
- 5–5 Emitter Feedback Biasing 97
- 5–6 Emitter Bypass Capacitor 100
- 5–7 Collector Feedback—CE Amplifier 103
- 5–8 Collector Feedback Input Impedance and Gain 105
- 5–9 Voltage Divider Feedback 107
  Problems 109

6 Field-Effect-Transistor Amplifiers 113

6–1 Introduction 113
6–2 JFET Theory of Operation 114
6–3 MOSFET Theory of Operation 120
6–4 Biasing Techniques—Depletion Mode FETs 122
6–5 Biasing—Enhancement-Mode MOSFETs 128
6–6 Negative Feedback Considerations 130
6–7 Source Follower 132
6–8 Conclusions 136
Problems 138

7 Power Amplifiers 141

7–1 Introduction 141
7–2 Heat Dissipation 143
7–3 Amplifier Distortion Effects 147
7–4 Class A Power Amplifiers 152
7–5 Class A Transformer Coupling 156
7–6 Push—Pull Operation 158
7–7 Additional Push—Pull Amplifier Consideration 162
7–8 Practical Push—Pull Amplifiers 165
Problems 168

8 Amplifier Frequency Response 171

8–1 Introduction 171
8–2 Capacitive Filters 172
8–3 Amplifier Performance at Low Frequencies 176
8–4 High–Frequency Response of the FET 180
8–5 High Frequency Response of the BJT 183
8–6 Review of BJT High Frequency Effects 188
8–7 Tuned Amplifiers—Introduction 190
8–8 Tuned Amplifiers 194
Problems 198

9 Multistage Amplifiers 201

9–1 Introduction 201
9–2 Darlington Compound 202
9–3 FET-BJT Combinations 206
9–4 Multistage Feedback 209
9–5 Three-Stage Amplifier 212
9–6 Audio-Power-Amplifier Design Example 216
     Problems 221

10 Oscillators 225

10–1 Introduction 225
10–2 Ringing in an LC Tank Circuit 226
10–3 Basic LC Oscillators 228
10–4 Hartly, Colpitts, and Clapp Oscillators 230
10–5 Crystal Oscillators 233
10–6 RC Phase-Shift Oscillator 235
10–7 Wein Bridge Oscillator 237
10–8 Negative-Resistance Oscillators 240
10–9 Parasitic Oscillations 243
      Problems 244

11 Linear Integrated Circuits 247

11–1 Introduction 247
11–2 Monolithic IC Fabrication 251
11–3 General LIC Considerations 253
11–4 Special Design Considerations 255
11–5 A General-Purpose LIC—The CA3035 262
11–6 LIC Power Amplifiers 267
      Problems 270

12 Linear Integrated Circuit Operational Amplifiers 273

12–1 Introduction 273
12–2 Ideal Operational Amplifier 274
12–3 Applications 278
12–4 Internal Circuitry 284
12–5 Frequency Considerations 286
12–6 Common–Mode Effects 289
12–7 The $\mu A741$ Operational Amplifier 290
Problems 297

13 Survey of Communication Systems 301

13–1 Introduction 301
13–2 Communication Systems and Noise Effects 303
13–3 Amplitude Modulation 306
13–4 Frequency Modulation 309
13–5 Transmission Methods—Antennas 312
13–6 Transmission Methods—Transmission Lines and Waveguides 316

Index 323

Preface

This textbook is written for electronic technology programs at community colleges and technical institutes. It also provides an up-to-date refresher course for working technicians and engineers. The mathematical treatment is light (simple algebra and trigonometry) with explanations made via logical reasoning rather than via sophisticated mathematical proofs.

This book may be used successfully by students who have completed a basic ac/dc course. For those students who have also completed a basic semiconductor devices course, Chapters 1 and 4 will be a review and can be covered briefly. If a basic devices course has not preceded this text, then an emphasis on Chapters 1 and 4 will be sufficient introduction to allow mastery of the ensuing topics. Both cases have been tried using the original manuscript of this text and have proven successful.

My reasons for writing this text are based on three arguments as outlined below:

1. Most textbooks have failed to keep pace with the dynamic changes in electronics today. In many cases the newest circuits and/or devices now in widespread use are omitted or only lip-service is provided by currently available textbooks.
2. All too often current textbooks analyze basic electronic circuits in such detail that reader comprehension is impaired. Fortunately,

there are suitable simplifications and approximations that predict circuit behavior without obscuring the reader's comprehension. This approach also yields surprisingly accurate results and is acceptable for all but the most exacting situations.
3. A textbook should be written so that the reader can understand it without constant assistance from a teacher. This book allows self-instruction for those not enrolled in a class, and in formal classroom situations it frees the instructor for laboratory work or supplemental efforts.

In conjunction with this text material I have, for three successive years, used a relatively unique laboratory approach. At the start of the course the student is given a box of electronic devices from which he is to design, construct, debug, and operate a complete electronic system, which must meet certain specifications including the following:

1. A regulated dc power supply
2. A small signal amplifier
3. A class AB push-pull power amplifier
4. A variable frequency oscillator

These four subsystems are tied together to form a complete electronic system that the student may keep upon completion of the course. The extreme success of this program is evidenced by our inability to keep students *out* of the laboratory at all hours of the day during the time span of this assignment. The students are highly motivated, and it is well known that efficient learning takes place under such conditions, apart from the practical value of this laboratory experience.

Chapters 2 and 3 on power supplies may be skipped or delayed without adversely affecting the student's progress in following material. However, because the subsequent chapters usually rely on ac to dc conversion as a source of power, I feel that it is logical if they are covered early in the course.

As a final note, I express appreciation to my wife for her patience and understanding during the preparation of this material and to Kathy Langworthy for many hours of typing the manuscript.

Gary M. Miller

*LINEAR CIRCUITS
FOR
ELECTRONICS
TECHNOLOGY*

1

Introductory Concepts

1-1 Introduction

In this book the basic circuits used in today's communication equipment are presented. These circuits are most often referred to as

1. Communication circuits.
2. Linear circuits.
3. Sinusoidal circuits.

Perhaps the most descriptive of these three is linear circuits, and hence it was chosen for use in the title of this textbook. A *linear circuit* may be defined as one in which the output signal is at all times in direct proportion to the input signal. The major topic of study in the book is amplifiers and their many forms. Amplifiers are the basis of every chapter, except for Chapters 2 and 3, in which dc power supplies are introduced, and Chapter 13, which is an introduction to communication systems.

Chapter 1 provides the fundamental concepts of *pn* junctions and transistors. An introduction to the concepts of amplification is also provided, and an important relationship between a transistor amplifier's parameters is developed.

At the very heart of your study of linear circuits is the transistor. This device has, in a little over two decades, caused a total transformation of the electronics industry. New products have been made possible and older ones made more compact, reliable, and economical. As a case in point, consider a portable *AM* radio receiver. In 1950, just before the commercial availability of transistors, typical vacuum-tube units had a cost of $30.00 and replacement battery packs sold for around $5.00. This radio was heavy and bulky compared to today's transistor receiver, which is pocket sized and sells for around $5.00 with a replacement battery at $0.29.

Since their introduction, transistors have been rapidly refined and developed to supplant their vacuum-tube counterparts in virtually all areas, except for those applications requiring *both* extremely high power outputs and high-frequency operation. A primary example of these applications is a radio transmitter with a power output of 1 kilowatt (kw) or more.

The characteristic of a transistor that makes it so important is its ability to convert a small signal into a similar but larger signal. This process and its many variations fall under the general category termed *amplification*. This characteristic is the basis of your study here. In the next section you will be introduced to the simple *pn* junction, which is more commonly referred to as a diode. With some additional theory, you will then move toward an understanding of transistor action in linear circuits.

1-2 pn *Junction Diode*

A *pn* junction is a combination of two types of semiconductor material. A semiconductor is a material whose electrical properties lie between those of an insulator and a conductor. The most often used semiconductors are either germanium or silicon. Crystals of either of these materials are modified to create either a *p*-type or an *n*-type semiconductor. This modification process is referred to as *doping*, and is accomplished by adding impurity atoms to pure germanium or silicon to increase the number of free electrons or holes. A *hole* is an electron vacancy in the outer shell of an atom. The *n*-type semiconductor is doped to produce an excess of free electrons. The *p*-type semiconductor is doped to produce an excess of holes (i.e., a deficiency of free electrons).

The *pn* junction exhibits the interesting and useful characteristic of offering a low resistance to current flow in one direction and a high resistance in the opposite direction. A *pn* diode and its symbol are shown in Fig. 1-1a. It is shown that conventional current can easily flow from the *p*-type to the *n*-type (Fig. 1-1b) material. Thus the direction of the arrowhead in a diode's schematic symbol functions as a memory aid for the direction of allowable conventional current flow. We shall use conventional current flow throughout this text. The student should be aware that the actual direction

pn *Junction Diode* 3

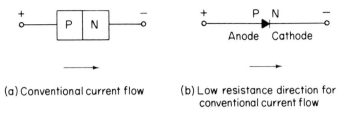

(a) Conventional current flow

(b) Low resistance direction for conventional current flow

FIG. 1-1. *pn* junction diode

of electron flow is in the opposite direction—from the *n*-type to the *p*-type material. This is logical since it is the *n*-type material that has been doped so as to have an excess of free electrons. When the diode is polarized for easy conduction (anode positive with respect to cathode), we say that the diode is *forward biased* (Fig. 1-2a).

Figure 1-2a illustrates a condition of *forward bias*. Forward bias results when the diode's *p*-type material is at a more positive level than the *n*-type material, which allows for easy conduction of current through the diode. Ideally, this easy conduction of current would mean zero voltage drop across the diode with the full battery voltage appearing across the resistor. As can be seen from Fig. 1-2a, however, a small voltage drop of 0.3 volt (V) for a germanium diode and 0.7 V for a silicon diode actually occurs. This difference of potential at the junction is called the *barrier potential*. Whenever the battery voltage is large enough to overcome this barrier potential, the diode presents a low resistance to the circuit and significant current flow may result. This is shown graphically by the current–voltage relationships in the first quadrant of Fig. 1-3 (positive current and voltage). The forward current is extremely low until the barrier potential is reached ($\simeq 0.3$ V in this case, indicating a germanium diode). As the forward voltage increases above the barrier potential, a rapid increase in current results. The diode is now presenting an extremely low resistance, and the current flow must be limited

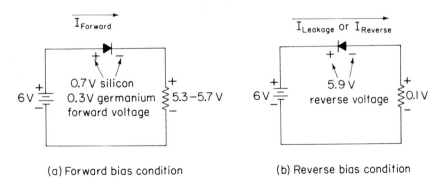

(a) Forward bias condition

(b) Reverse bias condition

FIG. 1-2. Diode bias conditions

4 *Introductory Concepts*

by external circuit resistance to prevent burnout of the diode. The power dissipated by a diode equals the product of the voltage across it and the current through it. In the case shown in Fig. 1-3, burnout occurs at 1 ampere (A) and about 0.5 V. Hence the power rating of this device is $\frac{1}{2}$ watt (W).

Referring back to Fig. 1-2a, the 6 V battery forward biases the diode with all the battery's voltage applied across the load, except for the diodes barrier potential of about 0.3 V if germanium, or 0.7 V if silicon. This then leaves either 5.7 or 5.3 V across the load when the diode is forward biased. In Fig. 1-2b the diode has been physically reversed, which results in a *reverse bias* on the diode. A reverse-biased diode has the *n*-type material (the cathode) at a more positive voltage level than the *p*-type material. The battery is now aiding the diode's barrier potential in preventing current flow, and only a very small reverse leakage current flows. This is shown graphically in the third quadrant of Fig. 1-3 (negative voltage and current). Notice the current remains at a very low level until the device's *breakdown voltage*, V_B, is exceeded. The breakdown voltage varies, depending on diode type, from a range of several volts up to 1000 V or more. If the breakdown voltage is exceeded, device burnout will once again occur, as shown in Fig. 1-3, if its power rating is exceeded. Notice that the reverse-biased condition of Fig. 1-2b results in almost all the battery's voltage being dropped across the diode with a very small voltage developed across the load resistor due to the small leakage current.

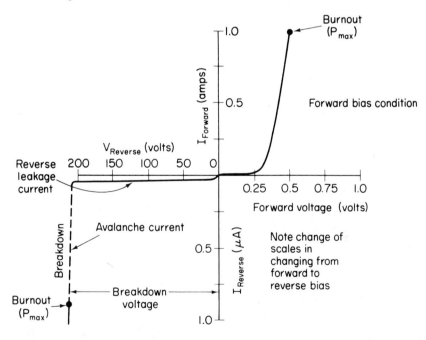

FIG. 1-3. Current–voltage relationships for a *pn* diode

Whenever a diode is heavily forward biased, the forward voltage drop across it is greater than just its knee voltage. This is shown in Fig. 1-3, where that particular diode drops 0.5 V with a forward current of 1 A. Thus the device is exhibiting a resistance that causes a voltage drop of 0.5 V minus the knee voltage of 0.3 V. This resistance is designated as the *bulk resistance*, r_B. An additional resistance is also present in a forward-biased diode. The *junction resistance*, r_j, must also be considered. It depends on the amount of forward dc current, and

$$r_j \simeq \frac{0.026 \text{ V}}{I_F} \tag{1-1}$$

where I_F is the dc forward current in amperes. The total ac resistance of a forward-biased diode is given by the sum of r_B and r_j, or

$$r_{ac} = r_B + r_j \tag{1-2}$$

A reverse-biased diode is not a perfect open circuit, as shown in Fig. 1-3, but exhibits some high value of *reverse resistance*, R_R. The reverse resistance of a diode can be approximated as simply its reverse voltage, V_R, divided by the reverse current, I_R. The manufacturer's data sheet usually supplies the necessary information to enable a calculation of R_R by providing a value of I_R at a specific value of V_R. Figure 1-4 summarizes resistance effects for both forward- and reverse-biased diodes by showing equivalent circuits for those two conditions.

EXAMPLE 1-1

A silicon diode has a forward voltage drop of 1 V when the forward current is 100 milliamperes (mA). It has a reverse current of 1 microampere (μA) for a reverse

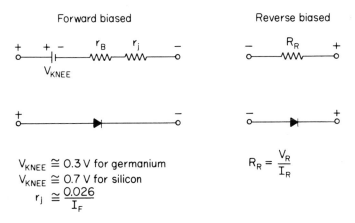

FIG. 1-4. Diode equivalent circuits

6 Introductory Concepts

voltage of 10 V. Calculate
  (a) The reverse resistance and the bulk resistance.
  (b) The total ac resistance at forward dc currents of 0.26, 2.6, 26, and 260 mA.

Solution:
(a) The reverse resistance is

$$R_R = \frac{V_R}{I_R} = \frac{10 \text{ V}}{1 \text{ }\mu\text{A}} = 10 \text{ megohms (M}\Omega\text{)}$$

The bulk resistance is

$$r_B = \frac{1 \text{ V} - 0.7 \text{ V}}{100 \text{ mA}} = \frac{0.3 \text{ V}}{0.1 \text{ A}} = 3 \text{ ohms (}\Omega\text{)}$$

(b) The total ac resistance equals the sum of the bulk resistance (calculated in part a) and the junction resistance:

$$r_{ac} = r_B + r_j \qquad (1\text{-}2)$$

The junction resistance varies from one value of dc current to another and must be calculated as

$$r_j \cong \frac{0.026}{I_F} \qquad (1\text{-}1)$$

when $I_F = 0.26$ mA, $\quad r_j = \dfrac{0.026 \text{ V}}{0.00026 \text{ A}} = 100 \text{ }\Omega$

$I_F = 2.6$ mA, $\quad r_j = \dfrac{0.026 \text{ V}}{0.0026 \text{ A}} = 10 \text{ }\Omega$

$I_F = 26$ mA, $\quad r_j = \dfrac{0.026 \text{ V}}{0.026 \text{ A}} = 1 \text{ }\Omega$

$I_F = 260$ mA, $\quad r_j = \dfrac{0.026 \text{ V}}{0.26 \text{ A}} = 0.1 \text{ }\Omega$

Since $r_{ac} = r_B + r_j$, then

when $I_F = 0.26$ mA, $\quad r_{ac} = 3 \text{ }\Omega + 100 \text{ }\Omega = 103 \text{ }\Omega$

$I_F = 2.6$ mA, $\quad r_{ac} = 3 \text{ }\Omega + 10 \text{ }\Omega = 13 \text{ }\Omega$

$I_F = 26$ mA, $\quad r_{ac} = 3 \text{ }\Omega + 1 \text{ }\Omega = 4 \text{ }\Omega$

$I_F = 260$ mA, $\quad r_{ac} = 3 \text{ }\Omega + 0.1 \text{ }\Omega = 3.1 \text{ }\Omega$

Thus the junction resistance decreases as the forward current increases. It is seen that as the forward dc current is increased, the ac resistance presented by the diode is decreased and approaches the bulk resistance of the diode.

1-3 Transistor Action

Transistors are made by forming two semiconductor junctions in close proximity. A *pnp* transistor is formed with *n*-type material between two pieces of *p*-type material, as illustrated in Fig. 1-5a. The larger region of

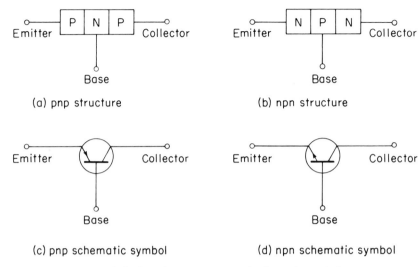

FIG. 1-5. Transistor structure and schematic symbols

p-type material is called the *collector* and the smaller p-type region the *emitter*. The n-type region in the center is termed the *base*. Figure 1-5b shows the *npn* counterpart, and beneath both structures is shown their corresponding schematic symbol. Notice that the arrowhead on the emitter lead changes direction depending on the transistor polarity—*npn* or *pnp*. As for the diode, the arrowhead points in the direction of easy conventional current flow from emitter to base.

We are now prepared to get to the very heart of transistor action. If the base–emitter junction is *forward* biased and the base–collector junction *reverse* biased, the transistor can be used to perform a very useful function. Figure 1-6 shows these bias conditions for a *pnp* transistor. A resistor has been included to limit current flow in the base circuit. The dc emitter current I_E can be calculated as 2.7 V minus the V_{BE} of 0.7 V (for silicon) divided by R_E or 2 V/1 kΩ = 2 mA. The base–collector bias is a 10 V reverse bias. A transistor connected as in Fig. 1-6 is known as the common-base (CB) configuration, since the base is common to both sides of the circuit. A set of output characteristic curves for the CB circuit is provided in Fig. 1-7.

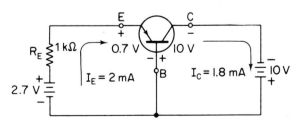

FIG. 1-6. Common-base transistor connection

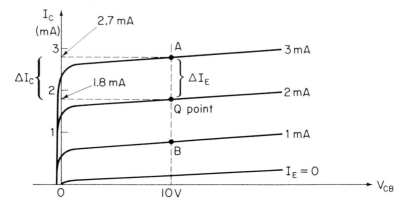

FIG. 1-7. Common-base characteristic curves

Notice that the curves relate information concerning the collector current as a function of applied V_{CB} for various values of input emitter current. From these curves the collector current of 1.8 mA can be determined from the 2 mA emitter current and 10 V reverse bias on the base–collector junction.

An important relationship exists between the collector and emitter currents in a CB circuit. Since the collector current is the output current and the emitter current is the input current, the ratio I_C/I_E is termed the dc CB forward current gain. It is given the symbol alpha (α) or h_{FB}. We shall use h_{FB} throughout the remainder of this book to designate forward current gain of the CB configuration:

$$h_{FB} = \alpha = \frac{I_C}{I_E} \tag{1-3}$$

The h_{FB} for the circuit of Fig. 1-6 is 1.8 mA/2.0 mA, or 0.9. Typical values of h_{FB} range from 0.9 to about 0.99, but are always less than 1.

If an *ac* signal were applied to the input side (emitter) of the circuit in Fig. 1-6, the ac value of CB forward current gain, h_{fb}, is of interest. It is defined as

$$h_{fb} = \left.\frac{\Delta I_C}{\Delta I_E}\right|_{V_{CB}=\text{constant}} \tag{1-4}$$

Figure 1-8 shows the CB circuit of Fig. 1-6 with an *ac* input current of 2 mA peak to peak (p-p). The 2-mA p-p current would cause the emitter current to vary around a specific dc bias level. That bias level (2 mA in this case) is called the *quiescent current* and is usually abbreviated as the *Q-point*. This is shown in Fig. 1-7, where the composite emitter current varies between 3 and 1 mA

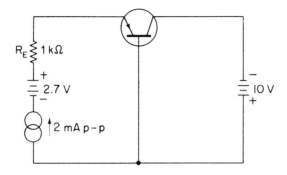

FIG. 1-8. Common-base circuit with ac input signal

around a 2 mA quiescent level (points A and B on Fig. 1-7). The ac current gain (h_{fb}) is calculated from the curves of Fig. 1-7 as follows:

$$h_{fb} = \left.\frac{\Delta I_C}{\Delta I_E}\right|_{V_{CB}=\text{constant}}$$
$$\simeq \left.\frac{2.7 \text{ mA} - 1.8 \text{ mA}}{3 \text{ mA} - 2 \text{ mA}}\right|_{V_{CB}=10 \text{ V}} \quad (1\text{-}4)$$
$$= 0.9$$

The h_{fb} was calculated between point A of Fig. 1-7 and the Q-point in this case, but could just as well have been calculated between point B and the Q-point. Note that in this instance the dc and ac current gains are equal. In actual practice the two are not usually exactly equal. A CB circuit provides a current gain of slightly less than 1 but is capable of substantial voltage gain, as will be shown in Chapter 4.

Another important transistor characteristic results when the emitter is made the common lead between input and output. As might be expected, this is termed the common-emitter (CE) configuration. The base is the input lead, whereas the output is at the collector. The CE forward current gain is the single most important transistor characteristic and is designated by the Greek letter beta (β) or h_{FE}. The latter term will be used throughout the remainder of this text. By definition, the dc CE forward current gain is

$$h_{FE} = \frac{I_C}{I_B} \quad (1\text{-}5)$$

Figure 1-9 shows a CE circuit with proper biasing, that is with the base–emitter junction forward biased and the base–collector junction reverse biased. Figure 1-10 shows a set of typical characteristics for the CE circuit of Fig. 1-9.

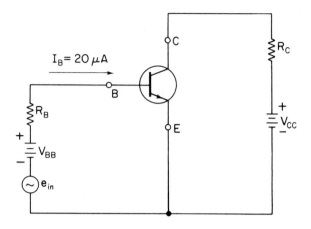

FIG. 1-9. Common-emitter circuit

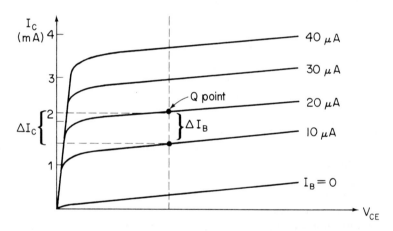

FIG. 1-10. Common-emitter characteristic curves

The Q-point conditions of $I_B = 20$ μA and $I_C = 2.2$ mA allow a calculation of h_{FE} as

$$h_{FE} = \frac{I_C}{I_B} \qquad (1\text{-}5)$$
$$= \frac{2.2 \text{ mA}}{20 \text{ μA}} = 110$$

As with h_{FB}, an ac current gain for the CE circuit is often used. It is defined similarly as

$$h_{fe} = \frac{\Delta I_C}{\Delta I_B}\bigg|_{V_{CE}=\text{constant}} \qquad (1\text{-}6)$$

If the ac input signal in Fig. 1-9, e_{in}, caused the base current to swing up to 30 μA and down to 10 μA around a 20 μA quiescent value, the ac forward current gain would be calculated with reference to the curves of Fig. 1-10 as

$$h_{fe} = \left.\frac{\Delta I_C}{\Delta I_B}\right|_{V_{CE}=\text{constant}}$$

$$= \frac{2.2 \text{ mA} - 1.5 \text{ mA}}{20 \text{ μA} - 10 \text{ μA}} \quad (1\text{-}6)$$

$$= \frac{0.7 \text{ mA}}{10 \text{ μA}} = 70$$

As can be seen, the ac current gain, h_{fe}, is significantly different from the dc current gain, h_{FE}. This is usually the case.

The reader should see at this point that (unlike the CB) the CE circuit is capable of providing a current gain, h_{fe}, to an ac input signal. It also provides significant voltage gain, as will be shown in Chapter 4. The ability to take a small signal and convert it into a large signal of the same shape is the most useful characteristic of the transistor. This process is known as *amplification*.

It can be easily shown by Kirchhoff's current law that a transistor's currents are related by

$$I_B = I_E - I_C \quad (1\text{-}7)$$

The base current is quite small with respect to the emitter and collector currents and is equal to their difference. For most calculations it is safe to assume that the emitter and collector currents are equal to one another. Equations (1-4), (1-6), and (1-7) may be combined to eliminate the transistor currents and show the relationship between h_{fb} and h_{fe}. This is left as an exercise for the student at the end of the chapter. The result is

$$h_{fe} = \frac{h_{fb}}{1 - h_{fb}} \quad (1\text{-}8)$$

As we shall see in future chapters, the term $(h_{fe} + 1)$ is often more useful than h_{fe}. As such, it is given the special designation, h'_{fe}. Thus

$$h'_{fe} = h_{fe} + 1 \quad (1\text{-}9)$$

Equations (1-8) and (1-9) may be rearranged to show that

$$h'_{fe} = \frac{h_{fe}}{h_{fb}} = \frac{1}{1 - h_{fb}} \quad (1\text{-}10)$$

The ac forms of h_{fe} and h_{fb} were used in Eqs. (1-8)–(1-10), but the equations are also valid for the dc forms (h_{FE} and h_{FB}).

EXAMPLE 1-2

A transistor has a dc base current of 1 mA and emitter current of 50 mA. Calculate I_C, h_{FB}, h_{FE}, and h'_{FE}.

Solution:

$$I_B = I_E - I_C$$
$$= I_E - I_B \qquad (1\text{-}7)$$
$$= 50 \text{ mA} - 1 \text{ mA}$$
$$= 49 \text{ mA}$$

$$h_{FB} = \frac{I_C}{I_E}$$
$$= \frac{49 \text{ mA}}{50 \text{ mA}} \qquad (1\text{-}3)$$
$$= 0.98$$

$$h_{FE} = \frac{h_{FB}}{1 - h_{FB}} = \frac{0.98}{1 - 0.98} \qquad (1\text{-}8)$$
$$= 49$$

or
$$h_{FE} = \frac{I_C}{I_B} \qquad (1\text{-}5)$$
$$= 49$$

$$h'_{FE} = h_{FE} + 1$$
$$= 49 + 1 \qquad (1\text{-}9)$$
$$= 50$$

or
$$h'_{FE} = \frac{1}{1 - h_{FB}}$$
$$= \frac{1}{1 - 0.98} = \frac{1}{0.02} \qquad (1\text{-}10)$$
$$= 50$$

EXAMPLE 1-3

The circuit from Example 1-2 has an ac input (base) current of 1 mA p-p, which results in an ac collector current of 40 mA p-p. Calculate h_{fe}, h'_{fe}, and h_{fb}.

Solution:

$$h_{fe} = \frac{\Delta I_C}{\Delta I_B}\bigg|_{V_{CE}=\text{constant}}$$
$$= \frac{40 \text{ mA}}{1 \text{ mA}} \qquad (1\text{-}6)$$
$$= 40$$

$$h'_{fe} = h_{fe} + 1$$
$$= 40 + 1 \qquad (1\text{-}9)$$
$$= 41$$

$$h'_{fe} = \frac{h_{fe}}{h_{fb}} \qquad (1\text{-}10)$$

$$h_{fb} = \frac{h_{fe}}{h'_{fe}} = \frac{40}{41}$$
$$\simeq 0.976$$

Notice that the ac gains are somewhat lower than the dc gains calculated in Example 1-2.

1-4 Amplification

Amplification has been defined as an enlargement of an input signal by an amplifier. If during this enlargement process the output signal does not maintain the exact same shape as the input, we say the output has been *distorted*. Distortion is illustrated in Fig. 1-11b. Note the distinction between an amplifier that just reproduces a signal with no amplification (Fig. 1-11a) and one that not only reproduces but amplifies as well (Fig. 1-11c).

Amplifiers may provide voltage, current, and power gains in varying

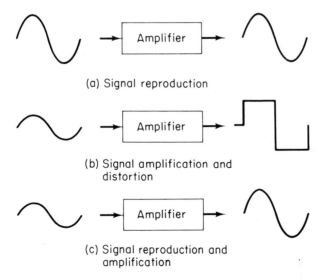

(a) Signal reproduction

(b) Signal amplification and distortion

(c) Signal reproduction and amplification

FIG. 1-11. Amplification and distortion

combinations. In almost all situations the gains we shall be studying are alternating current in nature, and hence the following definitions can be made:

$$\text{voltage gain} = G_v = \frac{\Delta e_{OUT}}{\Delta e_{IN}} \tag{1-11}$$

$$\text{current gain} = G_i = \frac{\Delta i_{OUT}}{\Delta i_{IN}} \tag{1-12}$$

$$\text{power gain} = G_p = \frac{\Delta P_{OUT}}{\Delta P_{IN}} = G_v \times G_i \tag{1-13}$$

These equations represent the gain of an *entire amplifier* stage, but if the gain of a device alone (such as a transistor) is considered, it is customary to refer to that as A_v instead of G_v, and A_i instead of G_i. This subtle differentiation is often confusing and requires careful attention by the student.

EXAMPLE 1-4

The ac input voltage to an amplifier stage is 1 V p-p and it causes an output variation of 20 V p-p. What is the voltage gain of this amplifier?

Solution:
Using Eq. (1-11), we have

$$G_v = \frac{\Delta e_{OUT}}{\Delta e_{IN}} = \frac{20\ V}{1\ V} = 20$$

EXAMPLE 1-5

If the same amplifier of Example 1-4 has an output current variation of from 10–50 mA for an input current variation of 1–2 mA, what are the current and power gains of this amplifier?

$$G_i = \frac{\Delta i_{OUT}}{\Delta i_{IN}} = \frac{50\ mA - 10\ mA}{2\ mA - 1\ mA}$$
$$= 40$$
$$G_p = G_v G_i = 20 \times 40$$
$$= 800$$

The term *impedance* is used to denote the effective ac resistance of an amplifier. The input impedance and the output impedance of an amplifier will be of great importance to us in our work with amplifiers. The input impedance Z_{in} is the ratio of input voltage change to the resulting input current change:

$$Z_{in} = \frac{\Delta e_{IN}}{\Delta i_{IN}} \tag{1-14}$$

Many texts refer to Z_{in} as simply R_{in}. The output impedance is similarly defined as

$$R_{out} = \frac{\Delta e_{out}}{Zi_{out}} \qquad (1\text{-}15)$$

EXAMPLE 1-6

Determine Z_{in} and R_{out} for the amplifier described in Examples 1-3 and 1-4.

Solution:

$$Z_{in} = \frac{\Delta e_{in}}{\Delta i_{in}} = \frac{1 \text{ V}}{2 \text{ mA} - 1 \text{ mA}} = 1 \text{ kilohm (k}\Omega)$$

$$R_{out} = \frac{\Delta e_{out}}{\Delta i_{out}} = \frac{20 \text{ V}}{50 \text{ mA} - 10 \text{ mA}} = 500 \text{ }\Omega$$

The input impedance gives us a measure of how much signal current the amplifier draws from its signal source.

Generally, a high value of Z_{in} is desirable. This seems logical, since many signal sources are not capable of delivering very much current, and a high input impedance will not draw a high value of current. The output impedance is a measure of how much current can be "extracted" from an amplifier to a load, and the lower R_{out} is, the greater this current can be. As such, a low value of R_{out} is generally desirable.

1-5 Basic Transistor Amplification

Transistors can be used to provide amplification because the amount of collector current flowing can be controlled by the very much smaller base current. Figure 1-12b shows a curve of collector current versus base current for a typical transistor; Fig. 1-12a shows the schematic of a transistor amplifier. Notice in Fig. 1-12b that there are two areas of nonlinearity, one for very low currents and one for high currents. In between is an area where $I_C = KI_B$, and this constant of proportionality K is equal to h_{FE}. Remember that $I_C = h_{FE}I_B$. For this particular transistor, h_{FE} is 20 in the linear region, since the collector current is always 20 times the base current. If a pure sinusoidal current were made to cause the base current to vary between 1 and 3 mA (i.e., a sinusoidal current of 2 mA p-p superimpressed on a dc level of 2 mA), then we would find a collector current variation of 20–60 mA. The collector current variation would also be a pure sinusoid, and hence is an exact replica of the input. This current *amplification* is illustrated in Fig. 1-13. Notice that both ac currents are riding on a dc level. These dc levels are referred to as the *bias levels* and are necessary to set up the proper operating point. For instance, without the 2 mA dc level in the base circuit provided by the battery V_{BB}, the negative excursion of the ac input signal would reverse bias the base–emitter junction, and transistor action would cease. These bias levels, as previously mentioned, are also termed the *quiescent operating point* (*Q*-point). The student should now realize that a *Q*-point (i.e., bias levels) must be initially set up in order to allow proper amplification of an ac signal. Without proper biasing, the transistor would be unable to amplify both the positive and negative extremes of the ac input signal, resulting in distortion.

One more important concept can be visualized with the help of Fig. 1-12 and 1-13. The input current *does not* supply current for the output cur-

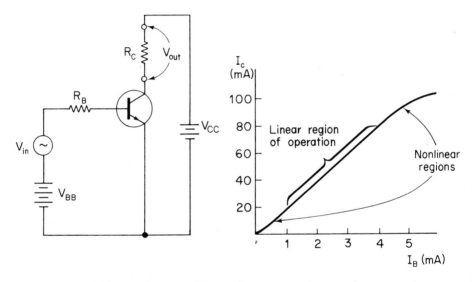

FIG. 1-12. Transistor amplifier; collector current versus base current characteristic curve of (a)

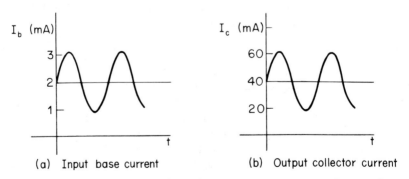

FIG. 1-13. Input and output current versus time for the amplifier illustrated in Fig. 1-12

rent! That is, only 2 mA p-p need be supplied by the input source to provide an output current of 40 mA p-p. One may wonder where all this output current comes from. The answer is simply that the battery in the collector circuit (V_{CC}) supplies this current. In other words, the small ac source impressed on the input of the transistor has resulted in a large ac draw on the battery in the collector circuit. The transistor here has acted as a "valve" in controlling the flow of electrons from the collector battery. The transistor has effectively taken a source of dc current (V_{CC}) and drawn from it a sinusoidal current! For the device to do this, however, it was driven by a much smaller sinusoidal current source into its base lead.

1-6 Gain–Impedance Relations

Amplifiers require four terminals for operation, since the input signal requires two leads and the output also requires two leads. Since transistors are only three-terminal devices, it is necessary to use one of the three leads as a common or shared lead between input and output. This presents no special problem, and often the common lead is grounded such that the input signal at point A in Fig. 1-14 is measured with respect to ground (terminal C) as is the output signal at B.

Transistor amplifiers exist in three possible configurations depending on which one of the transistor's three leads is used as the common terminal between input and output. The amplifier's designation is determined by the common terminal, as illustrated in Fig. 1-15. The performance of the amplifier is also determined by which lead is common. The CB and CE configurations have previously been demonstrated. Figure 1-15c shows the third possibility—the common collector (CC).

The gains and impedances of a transistor amplifier are interrelated, and these relationships can be expressed in a simple formula. Once derived and understood, this formula becomes an extremely useful and practical tool for evaluating and understanding amplifier operation. It is accurate for all transistor amplifiers regardless of configuration. It is not useful in evaluating

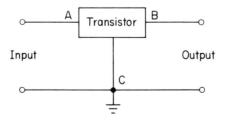

FIG. 1-14. Three-terminal amplifier

18 *Introductory Concepts*

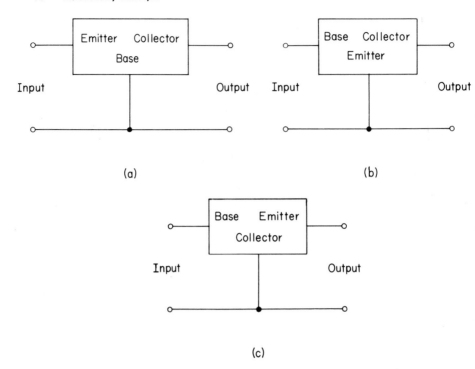

FIG. 1-15. Transistor amplifier classifications: (a) common base (CB); (b) common emitter (CE); (c) common collector (CC).

vacuum-tube or field-effect transistor (FET) amplifiers, and for this reason we shall call this formula the *transistor gain–impedance relationship*, hereafter referred to simply as the *TGIR*.

Derivation of the TGIR can be reasoned with the help of Fig. 1-16. From Ohm's law we know that

$$e_i = i_i R_{IN} \tag{1-16}$$

The output voltage then is

$$e_o = i_o R_L \tag{1-17}$$

Dividing Eq. (1-17) by (1-16),

$$\frac{e_o}{e_i} = \frac{i_o R_L}{i_i R_{IN}} \tag{1-18}$$

Since $G_v = e_o/e_i$ and $G_i = i_o/i_i$, we conclude that

$$G_v = G_i \frac{R_L}{R_{IN}} \tag{1-19}$$

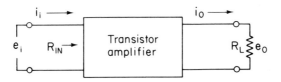

FIG. 1-16. Derivation of the TGIR

This is the TGIR. Its usefulness makes memorization worthwhile. As an aid to memory simply remember that GIRL's are in. ($G_i R_L$'s/R_{IN} where "are in" = R_{IN}). Interchanging G_v and G_i in Eq. (1-19) gives a useful variation of the TGIR:

$$G_i = G_v \frac{R_{IN}}{R_L} \qquad (1\text{-}20)$$

Note that this manipulation inverts the R_L/R_{IN} relationship of Eq. (1-19).

The voltage gain of a common-collector amplifier can be assumed to be 1, as we shall verify in Chapter 4. The input impedance of the CC amplifier will be shown to equal $h'_{fe} R_L$ in Chapter 4. Hence we can now use the TGIR variation, Eq. (1-20), to derive an expression for the current gain of a CC amplifier:

$$G_i = G_v \frac{R_{IN}}{R_L}$$
$$= 1 \frac{h'_{fe} R_L}{R_L}$$
$$= h'_{fe}$$

EXAMPLE 1-7

An amplifier has a G_i of 20, an input impedance of 1 kΩ and a load of 10 kΩ. Determine G_V.

Solution:

$$G_V = G_i \frac{R_L}{R_{IN}} \qquad (1\text{-}19)$$
$$= 20 \frac{10 \text{ k}\Omega}{1 \text{ k}\Omega} = 200$$

PROBLEMS

1. The diode in Fig. P1-1 is of the germanium variety.
   (a) Determine the voltage appearing across the AB terminals.

20 Introductory Concepts

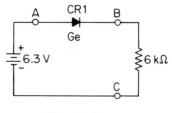

FIG. P1-1.

(b) Determine the voltage appearing across the *BC* terminals.
(c) Calculate the amount of current flow in this circuit.

2. Repeat Problem 1 with the battery terminals reversed.
3. Repeat Problem 1 with the germanium diode replaced by a silicon device.
4. A silicon diode is utilized in the circuit of Fig. P1-4. It has a leakage current of 0.1 mA.

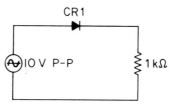

FIG. P1-4.

(a) Sketch the resistor's voltage waveform.
(b) What is the peak value of forward current flow?
(c) Calculate the power dissipation in the diode at the instant the input voltage is at its peak positive value.
(d) Calculate the power dissipation in the diode at the instant the input voltage is at its peak negative value.

5. A transistor has been properly biased (emitter diode forward biased and collector diode reverse biased) and has a dc emitter current of 10 mA and a base current of 0.2 mA.
   (a) Calculate the collector current.
   (b) Determine h_{FB}.
   (c) Determine h_{FE}.
   (d) Determine h'_{FE}.

6. A certain transistor has an I_C versus I_B curve as shown in Fig. P1-6.
   (a) Determine the range of base and collector currents through which this transistor will exhibit a linear relationship.
   (b) What is the name of the constant of proportionality between I_C and I_B?
   (c) Calculate the value of that constant for this transistor.

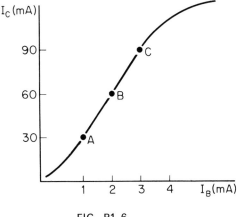

FIG. P1 6.

7. (a) An ac sinusoidal current of 1 mA p-p is applied to the base of the transistor utilized in Problem 6. The dc base current (Q-point) is at point B of Fig. P1-6. Sketch the resulting collector current.
   (b) The same current is now applied around a Q-point at point C of Fig. P1-6. Sketch the resulting collector current. (*Note:* The collector current is no longer a pure sinusoid and hence the amplifier has introduced distortion into the signal.)

8. (a) Which of the following characteristics did the amplifier exhibit in Problem 7a?
   (1) reproduction, (2) amplification, (3) reproduction and amplification.
   (b) Repeat for Problem 7b.

9. (a) Determine the voltage gain of an amplifier that has an input resistance of 1 kΩ and a load resistance of 5 kΩ. The current gain is 50.
   (b) If the load were doubled to 2.5 kΩ (i.e. I_L doubled), determine the new value of voltage gain.
   (c) Determine the power gain in parts a and b of this problem.

2

Rectification and Filters

2-1 Introduction

Virtually all pieces of electronic equipment require a dc source of energy for their operation as opposed to the ac power normally available from the power utilities. Batteries are a good source of this dc power for many applications—unfortunately, they are not satisfactory for many others because of size, weight, or cost considerations. The purpose of Chapters 2 and 3 is to acquaint you with the source of direct current used for the majority of electronic systems—the ac-to-dc power supply.

This power source is normally made up of the individual blocks illustrated in Fig. 2-1. The input at point A is normally 115 V ac, 60-hertz (Hz) line voltage, although it can be any convenient ac source. The transformer provides an efficient conversion of the input line voltage to the desired level. Hence the signal at point B will normally either be higher or lower than the input. It is usually lower, however, because of the low dc voltage level that

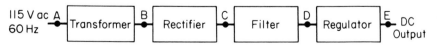

FIG. 2-1. Block diagram of a common dc power supply

most solid-state circuits require. The transformer also provides isolation from the power line—an important safety consideration. The explanation of the last three blocks is left to subsequent paragraphs.

The signal at point C in Fig. 2-1 is a dc signal. Hence all that is really required for an ac-to-dc power supply is a transformer and the rectifier. (In fact, even the transformer could be eliminated if no voltage transformation were required.) Thus we see that the filter and regulator are merely refinements of a dc power supply; but they are necessary for most applications.

2-2 Half-Wave Rectification

The most elementary method of rectification is the half-wave rectifier circuit shown in Fig. 2-2. When the transformer secondary voltage is on its positive half-cycle (E_{AB} positive), the diode $CR1$ is forward biased and hence presents a very low resistance to the voltage source. This means that almost all the secondary voltage is dropped across the load resistance R_L. The forward-biased diode will typically have a forward voltage drop of 0.5–1.0 V if it is silicon, or 0.2–0.6 V if germanium. To simplify analysis this voltage drop is often neglected, especially when the supply voltage is high, in which case the diode forward voltage drop will be a small percentage of the resulting output voltage.

Figure 2-2b illustrates the performance of a half-wave rectifier. Notice that the output V_{CD} is zero when the transformer voltage E_{AB} is negative. This occurs because the diode has become reverse biased (anode negative with respect to cathode). Recall that a reverse bias causes the diode to look like a very high resistance—ideally, an open circuit. Hence, by the voltage divider law, the voltage across R_L will approach zero. The average dc voltage (V_{dc}) is equal to 0.318 (0.318 = $1/\pi$) of the peak value. Many voltmeters

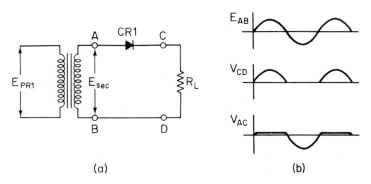

FIG. 2-2. (a) Half-wave rectifier circuit; (b) waveforms

are average reading devices and would thus register 0.318 of the peak voltage for a half-wave rectifier circuit. However, to calculate power, root-mean-square (rms) values must be used. The root-mean-square voltage for the half-wave rectifier circuit is 0.5 of the peak value:

$$V_{dc} = V_{average} = \frac{1}{\pi}E_P = 0.318E_P \qquad (2\text{-}1)$$

$$V_{rms} = 0.5E_P \qquad (2\text{-}2)$$

These two different methods of designating voltage can be a source of confusion. Fortunately, this is usually not a concern, however, because the filtered direct current we normally encounter has nearly equivalent root-mean-square and average values. The current obtained by dividing the load's average voltage by the load resistance is termed the average current, I_0. Hence

$$I_0 = I_{average} = \frac{V_{average}}{R_L} \qquad (2\text{-}3)$$

Referring back to Fig. 2-2b, notice that V_{AC} is the voltage across the diode $CR1$. When forward biased, there is a small voltage drop as previously mentioned. When reverse biased, however, the full input voltage is dropped across the diode. This is known as the peak reverse voltage (normally abbreviated PRV). Every diode has a maximum allowable PRV rating that should not be exceeded since device breakdown will usually occur. The diode voltage V_{AC} in Fig. 2-2b follows E_{AB} while the diode is reverse biased, since the diode has a very high resistance. Notice, also, that while $CR1$ is forward biased, the voltage across it (V_{AC}) is not zero but some small positive value. This is the diode's forward voltage drop, which is usually less than 1 V.

EXAMPLE 2-1

Referring to Fig. 2-3 the following data are given:
$E_{PR1} = 115$ V, 60 Hz.
$T1$ has a 10:1 turns ratio.

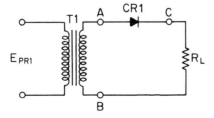

FIG. 2-3. Half-wave rectifier circuit for Example 2-1

26 Rectification and Filters

V_{forward} of $CR1 = 1$ V (when $CR1$ is forward biased).
R_{reverse} of $CR1 = 1$ MΩ (when $CR1$ is reverse biased).
$R_L = 100$ Ω.

Determine the following:
 (a) E_{AB} peak.
 (b) V_L (peak, average, and root mean square).
 (c) The PRV withstood by $CR1$.
 (d) The power dissipated in R_L.
 (e) The peak current through $CR1$.
 (f) The average current (I_0) through $CR1$.
 (g) The peak value of V_L during the negative half-cycle of E_{AB}.

Solution:

(a) Recall from your ac circuit theory that the root-mean-square voltage equals the peak voltage divided by $\sqrt{2}$; hence

$$E_{\text{peak}} = E_{\text{rms}} \sqrt{2}$$
$$E_{PR1} = 115 \text{ V ac}$$

Therefore

$$E_{PR1} \text{ peak} = 115 \text{ V} \times \sqrt{2} = 162 \text{ V}$$
$$E_{\text{sec}} \text{ peak} = \frac{162}{10} = 16.2 \text{ V} = E_{AB}$$

(b) $CR1$ has a forward voltage drop that we shall assume to be 1 V.

$$V_L \text{ peak} = E_{\text{sec}} \text{ peak} - V_{\text{forward}} \text{ of } CR1$$
$$= 16.2 \text{ V} - 1 \text{ V} = 15.2 \text{ V peak}$$

and

$$V_{L \text{ ave}} = V_L \text{ peak} \times 0.318 \quad \text{from Eq. (2-1)}$$
$$= 15.2 \text{ V} \times 0.318$$
$$= 4.8 \text{ V}$$
$$V_L \text{ rms} = V_L \text{ peak} \times 0.5 \quad \text{from Eq. (2-2)}$$
$$= 15.2 \text{ V} \times 0.5$$
$$= 7.6 \text{ V}$$

(c) The peak reverse voltage (PRV) of the diode equals the peak value of the transformer secondary voltage for a half-wave rectifier. Therefore

$$V_{\text{PRV}} = E_{\text{sec}} \text{ peak}$$
$$= 16.2 \text{ V}$$

(d) Recall that the power dissipated in a load equals the product of voltage times current, or I^2R or V^2/R. Remember that root-mean-square values of voltage and/or current must be utilized in making power calculations.

$$P = \frac{V^2 \text{ rms}}{R_L}$$

$$= \frac{7.6^2 \text{ V}}{100 \text{ }\Omega} = 0.58 \text{ W}$$

(e) $$I_{\text{peak}} = \frac{V_{L\text{ peak}}}{R_L}$$

$$= \frac{15.2 \text{ V}}{100 \text{ }\Omega} = 152 \text{ mA}$$

(f) $$I_{\text{ave}} = I_0 = \frac{V_{L\text{ ave}}}{R_L}$$

$$= \frac{4.8 \text{ V}}{100 \text{ }\Omega}$$

$$= 48 \text{ mA}$$

(g) To solve for V_L during the negative half-cycle, the voltage divider rule should be used. Figure 2-4a illustrates the voltage divider rule in general; part b applies this rule to the specific circuit in question:

$$V_L = 16.2 \text{ V} \times \frac{100 \text{ }\Omega}{100 \text{ }\Omega + 1 \text{ M}\Omega} \qquad \text{from Fig. 2-4b}$$

but since $100 \text{ }\Omega + 1 \text{ M}\Omega \cong 1 \text{ M}\Omega$,

$$V_L = \frac{16.2 \text{ V} \times 10 \text{ }\Omega}{10^6 \text{ }\Omega} = 16.2 \times 10^{-4} \text{ V}$$

$$= 0.00162 \text{ V}$$

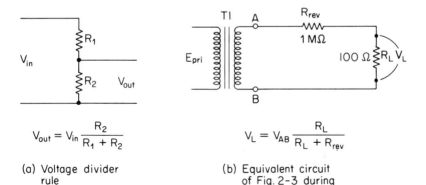

(a) Voltage divider rule

(b) Equivalent circuit of Fig. 2-3 during negative half-cycle

FIG. 2-4. Solution to Example 2-1

2-3 Full-Wave Rectification

A generally more useful and efficient method of converting alternating current to direct current is to recover both the positive and negative portions of the ac input signal. The two circuits used for this purpose are illustrated in Fig. 2-5. This is known as full-wave rectification in as much as the full input wave is utilized in the dc output.

The center-tap version shown in Fig. 2-5a uses two secondary windings connected "series aiding," as shown. If, at a given instant, the voltage polarities are as shown, then *CR*1 is forward biased and conducting, since its anode is positive with respect to its cathode, while *CR*2 is reverse biased and nonconducting. Thus *CR*1 only is supplying current to the load.

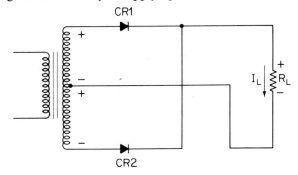

(a) Full-wave center-tap rectifier

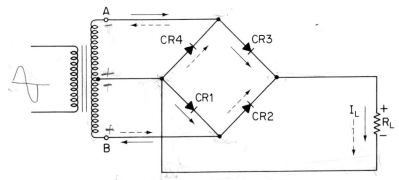

When V_{AB} is positive, solid arrows show current flow
When V_{AB} is negative, dotted arrows show current flow

(b) Full-wave bridge rectifier

FIG. 2-5. Full-wave rectifiers

This situation reverses itself when the polarity of the transformer's secondary voltage reverses on the next ac half-cycle. Then *CR*1 becomes reverse biased and *CR*2 forward biased, which means that *CR*2 is then supplying current to the load. Since each diode conducts only half the time on alternate half-cycles, a load current double the current rating of the diodes can be provided.

The input and output waveforms are shown in Fig. 2-6. Notice the effective doubling of frequency that takes place between the input and output. This is true since the period T of the output waveform is one half that of the ac input signal. Recall that frequency is inversely related to the period ($f = 1/T$).

The center-tap circuit was the most popular full-wave rectifier circuit until silicon diodes replaced vacuum diodes in most power supplies. The advent of these low-cost, highly reliable, and small-sized silicon diodes makes the bridge circuit the most popular version today. The reason for this is a reduction in the required transformer size for the same available output power as in the center-tap circuit. In the center-tap circuit, current is drawn through opposite halves of the transformer secondary on alternate half-cycles. On the other hand, the full-wave bridge circuit uses the entire transformer secondary during both half-cycles, allowing higher output power before core saturation than the center-tap circuit for equivalent-sized transformers. The obvious disadvantage of the bridge circuit is the need for twice as many rectifiers as the center-tapped version. However, with the availabil-

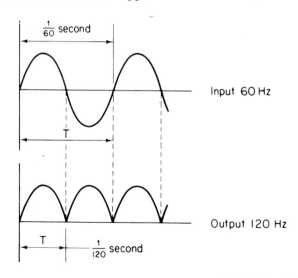

FIG. 2-6. Input (60 Hz) and subsequent output (120 Hz) for full-wave rectifiers

ity of low-cost silicon diodes, the full-wave bridge became more economical than the full-wave center-tap circuit, in spite of the need for twice as many diodes.

To analyze the performance of the full-wave bridge circuit, refer to Fig. 2-5b. The solid arrows show the current flow when the voltage V_{AB} is on the positive half-cycle; the dashed arrows denote current flow for the negative half-cycle. Notice that during the positive V_{AB} half-cycle $CR1$ and $CR3$ are conducting, and that during the negative half-cycle $CR2$ and $CR4$ conduct. This results in the full-wave voltage-output waveform shown in Fig. 2-6. Since the diodes conduct only half the time, the allowable average output current is double the average current ratings of the diodes. Notice that the peak value of the output voltage will be the peak transformer secondary voltage minus the forward drop of two diodes. Notice also that during the time $CR1$ and $CR3$ are forward biased (and hence conducting) $CR2$ and $CR4$ are reverse biased. As shown in Fig. 2-7, when V_{AB} is maximum positive, this voltage appears across the loop $ACDEB$. Writing the loop equation using Kirchhoff's voltage law, $V_{BA} + V_{AC} + V_{DE} + V_{EB} = 0$. Consider the instant that $V_{BA} = -10$ V as in Fig. 2-7. Then $V_{DE} = -10$ V approximately as a result of the conduction of $CR1$ and $CR3$. Hence

$$(-10 \text{ V}) + V_{AC} + (-10 \text{ V}) + V_{EB} = 0$$

and

$$V_{AC} + V_{EB} = 20 \text{ V}$$

Therefore, a total of 20 V is across the two diodes, and its seems logical to assume that they will split this voltage. Therefore, the reverse rating of each diode is equal to the peak input voltage of the circuit for a full-wave bridge circuit.

Full-wave rectification results in a root-mean-square output voltage of 0.707 E_{peak} and an average dc voltage of 0.636 E_{peak}. These results are summarized in Fig. 2-8 for full-wave and half-wave rectification.

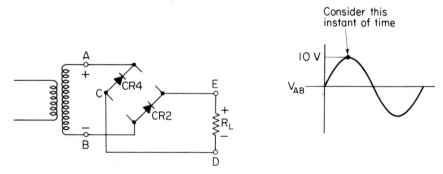

FIG. 2-7. Peak reverse voltage considerations—the reverse bias path

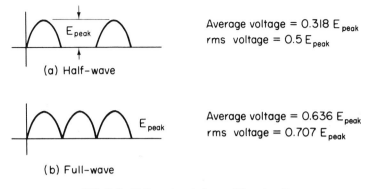

FIG. 2-8. Voltage levels for rectifier circuits

2-4 Filters for Power Supplies

The output of the various rectifier configurations we have considered is called *pulsating* direct current. It is pulsating because it pulsates from a zero level (ground) up to some peak value and back at a steady, repetitious rate. This type of output is sometimes useful to drive a dc motor, but not for sophisticated electronic circuits. As a matter of fact, these circuits usually require a very steady dc output that approaches the smoothness of a battery's output. A circuit that converts pulsating direct current into a very steady dc level is known as a *filter*, because it "filters" out the pulsations in the output. Refer to Fig. 2-9 for visualization of a filter's function.

Many types of filters have been devised over the years, using capacitors, inductors, resistors, and various combinations of the three. However, the capacitor alone has been almost exclusively used for the filters in today's solid-state power supplies, because inductors are extremely bulky and expensive compared to the lower-cost electrolytic capacitors that are available for today's low-voltage circuits. Consequently, we shall concentrate our study of filters in that area.

The amount of pulsation (usually termed the ac component) in a dc power supply output is called *ripple*. Keeping the ripple content low is the function of the filter. In practical applications, ripple is designated by

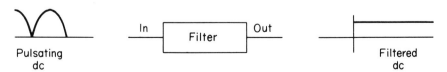

FIG. 2-9. Filter function

peak-to-peak volts or root-mean-square volts, and sometimes as a percentage of the output. Percentage ripple is defined as follows:

$$\% \text{ ripple} = \frac{\text{rms value of ripple}}{\text{dc output voltage}} \times 100\% \tag{2-4}$$

Notice that in Fig. 2-10 the power supply has a 10 V dc output with an ac component riding on the dc level. This ac component, the ripple, is designated in three different ways—all acceptable:

1. Peak-to-peak value of ac components of waveform.
2. Root-mean-square value of ac components in output waveform.
3. Per cent ripple [Eq. (2-4)].

Recall from your study of capacitance that a capacitor has the ability to store energy. The purpose of the capacitor filter is to smooth the rectifier output by receiving energy during peaks of rectifier output voltage for storage, and then releasing energy to the load during periods of low voltage input to the capacitor. This is illustrated for both the half-wave and full-wave condition in Fig. 2-11. Note that the capacitor's charge curve follows the sinusoidal rectifier output. This is possible because of the low resistance, and thus short time constant, of the charging circuit in Fig. 2-11a, which includes the transformer's secondary winding resistance (several ohms typically) and the low forward resistance of $CR1$ in parallel with the load resistance. You can better visualize this by referring to Fig. 2-12. Notice that for the typical values given in the figure, the total *charging* resistance ($CR1$ forward biased) is 4 Ω, whereas the discharge resistance ($CR1$ reverse biased) is 100 Ω. The discharge time constant is designed to be long enough to maintain the output voltage at as close to V_p as economically possible. The longer this RC discharge time constant (greater filter capacitance or higher load resistance), the better the filter action. The capacitor is thus "smoothing" the rectifier output, as pictured in Fig. 2-11d for a half-wave circuit and in Fig. 2-11f for a full-wave circuit.

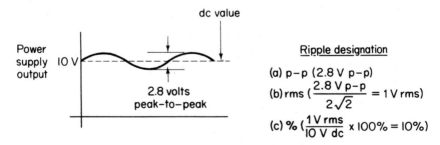

FIG. 2-10. Methods of ripple designation

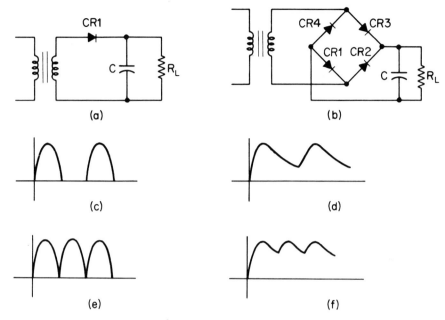

FIG. 2-11. Rectifier-filter circuits and waveforms: (a) half-wave circuit with filter; (b) full-wave circuit with filter; (c) half-wave output without filtering; (d) filtered half-wave output; (e) full-wave output without filtering; (f) filtered full-wave output.

It is noteworthy to consider the diode current flow in these circuits. With a capacitor filter, diode forward current flows during short intervals of time. These short conduction intervals correspond to the periods of time when the transformer secondary output voltage is greater than the capacitor voltage, which is also the load or output voltage. This is apparent, since the diode can only be conducting when the secondary voltage is greater than the load voltage. Thus the diode current is of a surging (temporary) nature, giving rise to an important rating of power rectifiers—surge current. This rating should not be exceeded in power-supply design. Given a rectifier's average current rating, the resulting surge current rating is adequate for most designs. The conditions the designer must be wary of are those circuits in which abnormally high values of filter capacitance are being utilized. This situation means that the charge current duty cycle is extremely short, and hence of a high amplitude.

It should be recognized that the magnitude of ripple is less in the full-wave circuit (Fig. 2-11f) than in the half-wave circuit (Fig. 2-11d) because of the shorter discharge time before the capacitor is reenergized by another pulse of current.

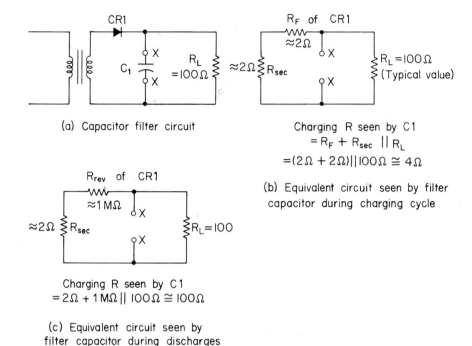

FIG. 2-12. Circuit analysis for half-wave rectifier

2-5 Practical Filter Design

One other criterion of ripple is useful to us when designing practical filter circuits—*ripple factor*—r, defined as follows:

$$r = \text{ripple factor} = \frac{\text{rms value of the ripple } (V_{ac})}{\text{average or dc component } (V_{dc})} \quad (2\text{-}5)$$

A reference to Eq. (2-4) shows that ripple factor is simply the percentage of ripple expressed as a fraction instead of a percentage (i.e., 10 per cent ripple is equivalent to a ripple factor of 0.1).

Having defined ripple factor, we are now in a position to use the chart illustrated in Fig. 2-13 for selection of the proper filter capacitor. The chart plots the ripple factor, r, versus the ωCR product. The ωCR product is simply the product of the angular velocity, filter capacitance, and load resistance. Note that as either ω or RC goes up, the ripple factor, r, and thus the amount of ripple, goes down. This indicates that more capacitance, lighter loads (higher load resistance), or a higher frequency will all serve to lower the rip-

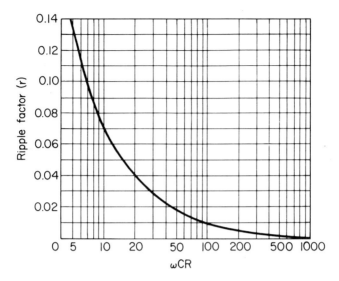

FIG. 2-13. Ripple factor versus ωCR for a full-wave circuit (From *Electronic Fundamentals and Applications,* John D. Ryder, 4th ed., Prentice-Hall, Inc., Englewood Cliffs, N.J., 1970.)

ple. Recall that ω (the angular velocity) $= 2\pi f$, and note from the chart that ωCR products over 100 do not appreciably increase filter performance. Thus increasing the filter capacitor's size beyond a certain limit will not appreciably improve the amount of ripple reduction. An interesting aspect of a full-wave rectifier is that it is a frequency doubler. This means that $\omega = 2\pi(120)$ for power supply calculations whenever a 60 Hz input signal is used.

Another power-supply design consideration is selection of the transformer's secondary voltage. In other words, what ac transformer output will provide the dc output that your application requires. Once again the ωCR product is of interest to us. If E_P is the peak value of the rectifier's output, then the curve in Fig. 2-14 gives us the ratio of the dc output voltage to E_P (the peak rectifier output voltage) versus ωCR for half-wave and full-wave circuits. Note that for the same ωCR product, the ratio is always higher for a full-wave (Fig. 2-14a) than a half-wave (Fig. 2-14b) configuration.

EXAMPLE 2-2

You are to design a power supply as shown in Fig. 2-15 operating from a 115 V, 60 Hz supply line. The load is 10 Ω and the load voltage is to be 10 V direct current with ripple of less than 1400 millivolts (mV) rms. Determine the
   (a) required value of filter capacitance and its voltage rating.
   (b) transformer secondary voltage required and VA rating.
   (c) diode's average current (I_0) and PRV.

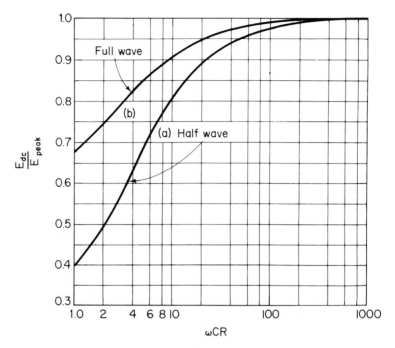

FIG. 2-14. Variation of E_{dc}/E_{peak} versus ωCR (From *Electronic Fundamentals and Applications,* John D. Ryder, 4th ed., Prentice-Hall, Inc., Englewood Cliffs, N.J., 1970.)

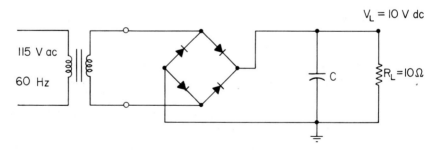

FIG. 2-15. Circuit for Example 2-2

Solution:
(a) First calculate ripple factor from Eq.(2-5):

$$r = \frac{1.4 \text{ V rms}}{10 \text{ V dc}}$$
$$= 0.14$$

Now from Fig. 2-13 the required ωCR product for a 0.14 ripple factor is approximately 5. Therefore,

$$\omega CR = 5$$

Remember that $\omega = 2\pi f$ and $f = 120$ Hz; therefore, $\omega CR = (2\pi \times 120) \times C \times 10\,\Omega = 5$. Solving for C,

$$C = \frac{5}{2\pi \times 120 \times 10\,\Omega}$$

$$= 0.664 \times 10^{-3} \text{ farad (F)}$$

$$= 0.664 \times 10^{3} \times 10^{-6} \text{ F}$$

$$= 664 \text{ microfarads } (\mu F)$$

The voltage rating of the capacitor should be at least equal to the output voltage of 10 V, but the optimum point of operation for an electrolytic capacitor for highest reliability is at about 80 per cent of its voltage rating. Hence a voltage rating of about 12 V would be best.

(b) Since $\omega CR = 5$, a reference to Fig. 2-14b gives an E_{dc}/E_{peak} ratio of 0.85. Therefore,

$$\frac{E_{dc}}{E_P} = 0.85$$

$$\frac{10}{E_P} = 0.85$$

$$E_P = \frac{10}{0.85}$$

$$= 11.8 \text{ V}$$

This means that the transformer secondary should be 11.8 V peak if the diodes have no forward voltage drop. However, in a full-wave bridge, we have the forward voltage drop of two series-connected diodes to contend with. Assuming 1 V forward drop per diode, the transformer should have an $11.8 \text{ V} + 2 \text{ V} = 13.8 \text{ V}$ peak output. Therefore, the root-mean-square transformer output voltage should be $13.8 \text{ V}/\sqrt{2} = 9.8 \text{ V}$.

One last rating of the transformer that should be considered is the power it must deliver. Essentially, all the secondary current reaches the load (except for minimal capacitor leakage current) and hence the secondary should have a 1 A rating. Occasionally, these transformers are rated by their volt–ampere (VA) product, which in this case would be about 10 V ac times 1 A or 10 VA.

(c) The load current $= 10 \text{ V}/10\,\Omega = 1$ A. In a full-wave bridge, any given diode is conducting only one half the time. Therefore, the average load current I_0 is $I_L/2$, or 0.5 A, and the diodes are all operating at 0.5 A.

The peak reverse voltage is 13.8 V since each diode must block the peak transformer secondary voltage. Hence PRV = 13.8 V.

2-6 Voltage Multipliers

A voltage multiplier circuit produces a greater dc output voltage than the ac input voltage to the rectifiers. Their use is justified when the power supply can be built at a lower cost than the equivalent circuit using a step-up transformer.

In the voltage doubler circuit of Fig. 2-16, assuming the upper ac terminal is positive, $CR1$ conducts and charges capacitor $C1$ to the peak value of the supply voltage (E_p). At the next ac half-cycle (upper transformer secondary terminal negative) $CR2$ conducts and charges capacitor $C2$ to the peak of that half-cycle. Since $C1$ and $C2$ are in series with each other and in parallel with R_L, the output voltage will approach the sum of the positive and negative peaks of the applied ac wave, or $2E_p$. Hence, if 115 V ac, 60 Hz were the ac input, the output would be $115 \times 2\sqrt{2} = 324$ V dc if there was no load connected to drain the peak charge from the capacitors. When used with a load, an output of less than $2E_p$ will result, as evidenced by the graph in Fig. 2-17, which shows that for ωCR products of less than 500 (extremely high) the dc output is less than twice the peak transformer secondary voltage.

With reference to Fig. 2-17, E_{DC} is the dc output voltage, E_{PEAK} is the peak value of the ac input voltage, C is the value of one of the two equal valued capacitors in the circuit, ω is the angular velocity of the ac input voltage, and R is the load resistance.

Large values of capacitance or very light loads are required for voltage multipliers to be effective. In theory, there is no limit to the amount of voltage multiplication that can be obtained. Voltage triplers and quadruplers are fairly common, but practical considerations limit additional multipli-

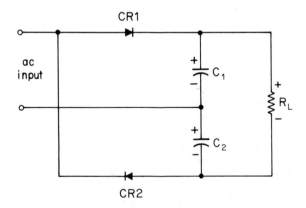

FIG. 2-16. Voltage-doubler circuit

cations. The amount of capacitance becomes unduly large to maintain the desired dc output voltage for anything except extremely light loads. Figure 2-18 illustrates a voltage-quadrupler circuit. Explanation of this circuit is left as an exercise for the student in the problems at the end of the chapter.

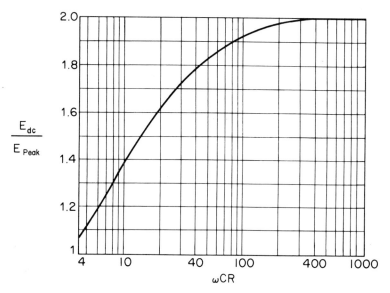

FIG. 2-17. E_{dc}/E_{peak} versus ωCR for the voltage doubler (From *Electronic Fundamentals and Applications,* John D. Ryder, 4th ed., Prentice-Hall, Inc., Englewood Cliffs, N.J., 1970.)

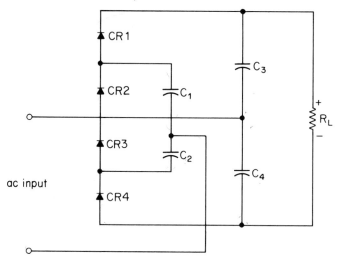

FIG. 2-18. Voltage-quadrupler circuit

PROBLEMS

1. (a) A 10 V ac voltage is applied to a silicon half-wave rectifier. Determine the average and root-mean-square output voltage. Include diode forward voltage drop effects.
   (b) Repeat part a if 2 V ac were the input voltage.
   (c) Repeat parts a and b but neglect the forward voltage drop of $CR1$.
   (d) Calculate the percentage of error introduced in both cases when the forward voltage drop is neglected.

2. A half-wave rectifier circuit is to provide an average voltage of 50 V dc.
   (a) Determine the peak value of the output voltage.
   (b) Calculate the root-mean-square voltage across the transformer secondary.
   (c) Calculate the root-mean-square output voltage.

3. A transformer with a 5:1 turns ratio has an input of 115 V ac. It drives a full-wave bridge rectifier and a 50 Ω load. Determine
   (a) peak transformer secondary voltage.
   (b) V_L (peak, average, and root mean square).
   (c) PRV withstood by the diodes.
   (d) peak diode current.
   (e) load power.
   (f) average diode current.

4. Repeat Problem 1 for a full-wave bridge circuit.

5. Determine V_{dc} across R_L for the doubler circuit of Fig. 2-16 if $E_{ac} = 115$ V, 60 Hz, and $C_1 = C_2 = 200$ μF, and $R_L = 100$ Ω.

6. Why is the full-wave circuit sometimes called a frequency doubler?

7. Explain why a full-wave bridge (FWBR) circuit enables use of a less costly transformer (less core iron) than the full-wave center-tap (FWCT) circuit.

8. Demonstrate that the required PRV ratings for both FWBR and FWCT circuits are equal to the peak value of the transformer secondary voltage.

9. Explain why the current flow through a rectifier diode, when used with a filter capacitor, is of a "peaky" nature.

10. Why are the root-mean-square and average values of output voltage in a well-filtered power supply nearly equivalent?

11. Explain the operation of the voltage quadrupler in Fig. 2-18.

12. Explain why the peak output voltage of a FWBR circuit will be the peak transformer secondary voltage minus the forward drop of two diodes.

3

Electronic Regulators

3-1 Introduction

The power supplies that we have thus far discussed are of the *unregulated* variety. A *regulated* power supply has more than just the basic transformer, rectifiers, and filter discussed in Chapter 2. Does a regulated power supply have a means of keeping the dc output voltage of an unregulated power supply constant

1. when the ac input voltage to the transformer varies (variations in the 115-V, 60-Hz line from 100–130 V are not uncommon)?
2. when the load is varied?

The answers to the above are fairly obvious—the dc output voltage still changes in a predictable fashion in both cases, but to a much lesser extent.

The term that gives an indication of the amount of this deviation for load changes is *percentage load regulation*, sometimes referred to simply as the power supply's *load regulation:*

$$\% \text{ load regulation} = \frac{V_{NL} - V_{FL}}{V_{FL}} \times 100\% \qquad (3\text{-}1)$$

where V_{FL} and V_{NL} refer to the power supply output under full-load and no-load conditions, respectively.

The term that gives an indication of the deviation for input line changes is the *percentage line regulation* or just *line regulation:*

$$\% \text{ line regulation} = \frac{(V_{high} - V_{low}) \times 100\%}{V_{high}} \quad (3\text{-}2)$$

where V_{high} refers to the dc output at the highest specified input voltage and V_{low} refers to the dc output at the lowest specified input voltage.

The term that predicts a supply's output variations with respect to both line and load variations is the *percentage regulation* or simply the *regulation:*

$$\% \text{ regulation} = \frac{(V_{max} - V_{min}) \times 100\%}{V_{max}} \quad (3\text{-}3)$$

where V_{max} is the maximum dc output voltage for variations in both line and load and normally occurs when the ac input is at its maximum value and when the load is minimum (R_L maximum). V_{min} is the minimum dc output voltage and normally occurs under $V_{IN\ min}$ and $R_{L\ min}$ conditions.

For highly regulated power supplies (i.e., very small dc output changes), another designator commonly replaces percentage regulation. Instead of an unwieldy regulation specification, such as 0.005 per cent, it is more convenient to simply say ± 0.25 mV. Thus a 10 V dc power supply with 0.005 per cent regulation could also be designated by the ± 0.25 mV regulation expression. Notice that 0.005 per cent of 10 V is 0.5 mV:

$$0.005\% \times \frac{1}{100\%} \times 10\ V = 0.5\ mV$$

The term 0.005 per cent in this case indicates a 0.5 mV envelope within which this power supply will remain for all line and load variations, as does the term ± 0.25 mV.

An unregulated power supply's output will vary directly with the input line voltage variation. If the line voltage changes from 115 V by 10 per cent to 126.5 V, then the dc output will also change by 10 per cent. For instance a 10 V dc power supply would go up to 11 V dc. This 10 per cent load voltage change due to a line change is fairly substantial and should give the reader an indication of the need for some form of voltage regulator for most electronic systems.

3-2 Ideal Voltage Source

The ideal voltage source has an output impedance, z_o, of 0 Ω. Then, regardless of the load it delivers power to, the voltage across that load would remain constant. In other words, a 10 V source with $z_o = 0$ Ω delivers 10 V to any load from an open circuit to a short circuit. This effect is illustrated in Fig. 3-1, where a 10 V battery with zero impedance is shown delivering 10 V to any variable load, R_L. Unfortunately, all practical sources of voltage always have some internal impedance (often called output impedance) associated with them, which will be designated as z_o. Figure 3-2 shows such a

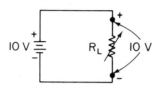

FIG. 3-1. Ideal voltage source and load

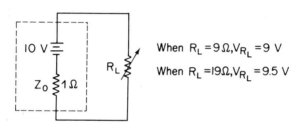

FIG. 3-2. Nonideal voltage and load

source, a 10 V battery with a z_o of 1 Ω. If R_L were 9 Ω, then the voltage across it by the voltage divider rule would be

$$V_{R_L} = 10 \text{ V}\left(\frac{9 \text{ Ω}}{1 \text{ Ω} + 9 \text{ Ω}}\right) = 9 \text{ V}$$

If the load were changed to 19 Ω, the load voltage would increase to

$$V_{R_L} = 10 \text{ V}\left(\frac{19 \text{ Ω}}{1 \text{ Ω} + 19 \text{ Ω}}\right) = 9.5 \text{ V}$$

Thus it is seen that the voltage delivered to a load from a nonideal voltage source is dependent on the size of the load. The more current drawn from a practical voltage source, the lower will be the voltage at the load due to the increased voltage drop across the source's internal output impedance z_o.

If z_o is not known for a voltage source, it can be determined by measuring the load voltage change for a known value of load current change. Then

$$z_o = \frac{\Delta V_L}{\Delta I_L} \tag{3-4}$$

where ΔV_L is the load voltage change resulting from a load current change, ΔI_L.

EXAMPLE 3-1

In adding a 1 A load to an existing 1 A load, a power supply's output voltage drops from 10.5 to 10 V. Calculate z_o, the output impedance of this power supply.

Solution:

$$\text{output impedance} = z_o = \frac{\Delta V_L}{\Delta I_L}$$

$$z_o = \frac{10.5 \text{ V} - 10 \text{ V}}{2 \text{ A} - 1 \text{ A}} = \frac{0.5 \text{ V}}{1 \text{ A}} = 0.5 \, \Omega$$

So it is seen that a power supply not only acts as a voltage source but also includes an output impedance. In Example 3-1 we could also calculate the output voltage under a no-load condition. If the output impedance, z_o, is constant throughout a range of 0 to 2 A load current, then the output voltage at no load could be calculated from Eq. (3-4), or

$$z_o = \frac{\Delta V_L}{\Delta I_L}$$

$$0.5 \, \Omega = \frac{V_{NL} - 10.5 \text{ V}}{1 \text{ A}} \quad \text{or} \quad 0.5 \, \Omega = \frac{V_{NL} - 10 \text{ V}}{2 \text{ A}} \tag{3-4}$$

$$V_{NL} = 1 \text{ A} \times 0.5 \, \Omega + 10.5 \text{ V} \quad \text{or} \quad V_{NL} = 2 \text{ A} \times 0.5 \, \Omega + 10.0 \text{ V}$$
$$= 11 \text{ V}$$

It can be seen if z_o went up, the output change would be even greater for a given load change. Hence the better regulated supplies have a very low output impedance. An equivalent circuit for the power supply just discussed is illustrated in Fig. 3-3.

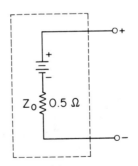

FIG. 3-3. Power supply equivalent circuit

The function of an electronic regulator is to take the z_o of the non-regulated supply and reduce it to a very low level. The perfect voltage source would have a z_o of zero ohms since, then, regardless of current draw, the output voltage would remain constant. Normally, the electronic regulator also serves to reduce the ripple at the same time that it is keeping the output at a relatively constant level. This will be demonstrated in a subsequent section.

The output impedance of a power supply for slowly changing loads is often referred to as its output resistance. For rapidly changing load levels—the normal situation in electronic systems—this static or dc output resistance is usually not valid. The output impedance varies with the frequency of the load change. This change with frequency occurs because of the reactance in the output of the power supply. Figure 3-4 illustrates a curve of z_o versus frequency for a high-quality supply. When specifying power supplies, this is an important criterion. The output impedance versus frequency curves are simply a way of stating the load regulation for a spectrum of frequencies.

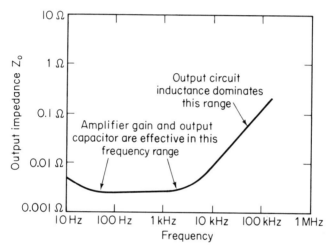

FIG. 3-4. Graph of Z_o versus frequency of load change for a well-regulated supply (Courtesy of Motorola Semiconductor Products, Inc.)

EXAMPLE 3-2

A dc power supply is delivering 10 V (nominally) to a load that is varying sinusoidally between $\frac{1}{2}$ and 1 A at a rate of 10 kHz. The graph in Fig. 3-4 accurately predicts its output impedance versus frequency characteristics. Determine the fluctuation in the output voltage caused by this periodic load change.

Solution:
The output impedance is noted to be 0.01 Ω at 10 kHz from Fig. 3-4:

$$z_o = \frac{\Delta V_L}{\Delta I_L}$$

$$\begin{aligned}
\Delta V_L &= \Delta I_L \times z_o \\
&= (1 \text{ A} - 0.5 \text{ A}) \times 0.01 \text{ Ω} \\
&= 0.005 \text{ V} = 5 \text{ mV}
\end{aligned}$$

Therefore, the output voltage will have a 5 mV p-p fluctuation at a rate of 10 kHz.

It is now an easy matter to convert this to a load regulation specification of 10 mV (5 mV for $\frac{1}{2}$ A changes and hence 10 mV for a full 1 A change) for 0 to 1 A load changes at frequencies up to 10 kHz. This is now a *dynamic* regulation specification, since it covers dc changes as well as higher-frequency (dynamic) changes. Notice that at frequencies above 10 kHz the z_o of this supply goes up, and therefore its dynamic load regulation would become poorer above 10 kHz, as shown in Fig. 3-4.

3-3 Zener Diode Shunt Regulators

Now that we have briefly talked about the characteristics of electronic regulators in general, we shall study the internal details of the elementary forms. A zener diode is similar to a normal diode except that its reverse breakdown voltage is put to practical use. In addition, the point of reverse breakdown is carefully controlled in the manufacturing process, and this zener voltage is available anywhere from 2 or 3 V up to 200 V. Figure 3-5 illustrates the zener diode characteristic curve with the pertinent currents and voltages labeled. I_{ZK} is the minimum reverse diode current to put the diode into its regulating region, and I_{ZM} is the maximum current the diode can withstand without exceeding its power rating. Notice that between I_{ZK} and I_{ZM} the zener voltage changes only slightly. The slope of the line between I_{ZK} and I_{ZM} (slope $= \Delta V/\Delta I$) is known as Z_z, the *zener impedance*.

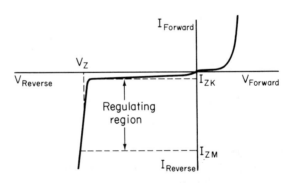

FIG. 3-5. Zener diode characteristic curve

By operating a zener diode between I_{ZK} and I_{ZM}, the voltage across the diode will remain relatively constant. This principle is put to practical use in the circuit shown in Fig. 3-6. An unregulated source of dc voltage from 20–30 V is applied to the circuit. The polarity is such as to reverse bias the zener diode and, it is hoped, to supply enough reverse current so that I_{ZK} will be surpassed and hence will put the diode into the *regulating region* shown in Fig. 3-5. The output voltage will then remain relatively constant, even as the input varies from 20–30 V. The zener diode is thus regulating the

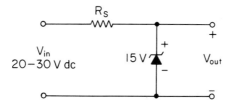

FIG. 3-6. Simple shunt regulator

output voltage and is in shunt (parallel) with the output and the circuit, and is therefore termed a *shunt regulator*.

The value of R_S in Fig. 3-6 must be properly selected so as to always provide enough current to keep the zener regulating and to supply the necessary load current. The necessary reverse zener current is I_{ZK}, and the load is usually variable between certain specific limits. If the load were to vary from 0–100 mA and I_{ZK} for the zener diode were 10 mA, it would be possible to calculate a suitable value for R_S. The current through R_S will be a minimum when the voltage across R_S is minimum. Therefore, R_S must be small enough to supply the maximum load current of 100 mA and the zener diode's I_{ZK} in order to allow proper circuit operation under all conditions. Thus

$$R_S = \frac{V_{RS\ min}}{I_{RS\ max}} \tag{3-5}$$

$$= \frac{V_{IN\ min} - V_Z}{I_{RS\ max}} = \frac{20\ V - 15\ V}{100\ mA + 10\ mA}$$

$$= \frac{5\ V}{110\ mA} = 45.5\ \Omega$$

If an R_S any greater than 45.5 Ω were used, the zener would not be provided with enough reverse current (I_{ZK}) when the unregulated input is at its minimum value (20 V) and the load is drawing its maximum amount of current (100 mA). As a safety factor, it is wise to select R_S slightly smaller than the exact calculated value.

Once the proper value of R_S has been selected for a shunt regulator, it is necessary to ensure that the zener diode's power rating is not exceeded when R_S supplies it with a higher current than it does at the minimum input voltage and/or maximum load current conditions. If the input voltage increases from the minimum value of 20 V, the voltage across R_S increases, since the zener voltage remains relatively constant. Similarly, if the load changed so as to draw less than 100 mA, that drop in load current would be forced through the zener diode. Thus we see that the maximum power dissipation in the zener diode of a shunt regulator occurs when the input goes to its highest possible value and when the load current is at its minimum value. For the circuit of Fig. 3-6 then, with an R_S of 45.5 Ω, the maximum zener current can be calculated as

$$I_{Z\,max} = \frac{V_{RS\,max}}{R_S} - I_{L\,min} \tag{3-6}$$

Since $V_{RS\,max}$ is 30 V − 15 V and $I_{L\,min}$ was specified as zero, we have

$$I_{Z\,max} = \frac{30\,\text{V} - 15\,\text{V}}{45.5\,\Omega} - 0\,\text{mA}$$

$$= 330\,\text{mA}$$

The power dissipated by a zener diode is equal to the product of its voltage and current:

$$P_Z = V_Z I_Z \tag{3-7}$$

Thus the maximum power dissipated by the zener diode in the circuit of Fig. 3-6 occurs when $V_{IN} = 30$ V and gives a V_Z of 30 V − 15 V:

$$P_Z = 15\,\text{V} \times 330\,\text{mA}$$
$$= 4.95\,\text{W}$$

EXAMPLE 3-3

Calculate an appropriate value for R_S and the required power rating of $CR1$ for the circuit shown in Fig. 3-7. The input varies from 15–20 V and $I_{ZK} = 10$ mA.

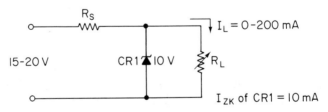

FIG. 3-7. Example 3-3

Solution:
Noting from Fig. 3-7 that the load current can vary from 0–200 mA, the maximum allowable value of R_S can be calculated from Eq.(3-5):

$$R_S = \frac{V_{RS\,min}}{I_{RS\,max}}$$
$$= \frac{15\,\text{V} - 10\,\text{V}}{200\,\text{mA} + 10\,\text{mA}} = \frac{5\,\text{V}}{210\,\text{mA}} \tag{3-5}$$
$$= 23.8\,\Omega$$

As a safety factor, select $R_S = 22\,\Omega$, a standard value of resistance slightly lower than the calculated value of R_S. The maximum current through the zener will occur when the load is minimum (0 mA) and the input is maximum (20 V). From Eq. (3-6),

$$I_{Z \text{ max}} = \frac{V_{RS \text{ max}}}{R_S} - I_{L \text{ min}}$$

$$= \frac{20 \text{ V} - 10 \text{ V}}{22 \text{ }\Omega} - 0 \text{ mA} \qquad (3\text{-}6)$$

$$= 455 \text{ mA}$$

Thus the required zener power rating can be calculated from Eq. (3-7) as

$$P_Z = V_Z I_Z$$
$$= 10 \text{ V} \times 455 \text{ mA}$$
$$= 4.55 \text{ W}$$

Although shunt regulators are effective in maintaining a constant load voltage as the source voltage changes and/or the load changes, they are relatively inefficient. Large amount of power are wasted in the zener and R_S in comparison with the amount of power actually delivered to the load. The series regulator covered in the next section overcomes this difficulty.

3-4 Series Regulators

In contrast to the shunt regulator, the control element of the series regulator is in series with the load. The block diagram shown in Fig. 3-8 is useful in analyzing the performance of such a circuit. The first two blocks are the familiar elements of a basic unregulated power supply. The series pass element is the control element—the element that turns the "electron spigot" up and down to maintain a constant voltage at the load, which may be changing, or to compensate for changes in the input voltage. It is normally a power transistor that performs this control function. The reference element is a constant voltage device that compares the output voltage with

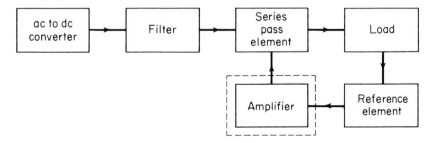

FIG. 3-8. Block diagram for a series regulator

itself and then sends out an "error" signal to the control element when necessary. This error signal is passed on to an amplifier, which then relays it to the series pass element. In the simpler circuits, a separate amplifier is not used—the reference element feeds the "error" signal directly to the series pass element, which also functions as the amplifier.

What has just been described is a *feedback control* system. This is a system that somehow *senses* its own output and, if the output is not correct, sends an error signal to the amplifier to be amplified and then to the control element to correct this condition. In our particular system the output voltage might instantaneously change from the desired level, which, when compared with the reference element, results in an error signal. Figure 3-9 illustrates a simple but effective series regulator circuit.

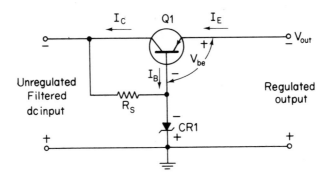

FIG. 3-9. Simple series regulator circuit

With reference to Fig. 3-9, the zener diode $CR1$ is the reference element. $Q1$ is the series pass element and also the amplifier in this simple series regulator circuit.

Referring to Fig. 3-10, we see that the voltages V_{BE}, V_{CR1}, and V_{out} form a closed loop. Kirchhoff's voltage law states that the algebraic sum of the voltages in a closed loop is zero. Therefore, it stands to reason that if V_{BE} and V_{CR1} are relatively constant, as indeed they should be in a properly working circuit, V_{out} should also be a constant voltage. This is obviously the desired result for a voltage regulator, and that is exactly how this circuit works, at least in very simple terms.

Let us now take a look at the circuit operation in a different way and in greater detail. R_S supplies current for the base of $Q1$ and for $CR1$ to keep it operating on the zener breakdown area (regulating region). Care must be taken that R_S has a low enough value to be able to always supply the needs of these two devices. The worst-case condition in this respect occurs at the minimum input voltage and maximum output current. This is because the voltage across R_S is then minimum, making I_{RS} minimum, and I_E and I_C of

Series Regulators   51

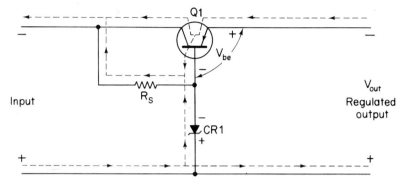

FIG. 3-10. Series regulator detail showing voltage polarities and current flow

$Q1$ are maximum, which means that R_S must then supply the highest value of base current. The zener diode "absorbs" any of the excess current from R_S not needed for the transistor's base, but requires some minimum value to keep it operating in its regulating region (I_{ZK}—refer to Fig. 3-5).

If, for instance, the load on this power supply were suddenly doubled (i.e., I_L doubled), the output voltage would instantaneously tend to drop, because the unregulated input would drop as more current is drawn from it. However, the constant voltage effect of the zener diode causes a larger forward bias (V_{BE}) on $Q1$; hence I_B of $Q1$ will increase and cause a greater I_C flow and thus a greater I_E flow. This increase in bias occurs because the load voltage has instantaneously dropped while the zener voltage will, it is hoped, remain constant. Since V_{BE} is the only other voltage drop in the loop with V_Z and V_L, it stands to reason that it must increase instantaneously. This results in a larger forward bias and hence greater load current. This greater load current then tends to bring the load voltage back up to its original value. Thus $Q1$ is acting as a "valve" in controlling current flow in order to obtain a constant voltage output. An increase in output voltage due to a change to a lighter load is compensated for in a similar but naturally reverse fashion. This circuit also compensates for changes in input voltage (line regulation) in a similar manner. A circuit of this type can typically offer a 5 per cent regulation figure.

The speed with which this regulator "reacts" to input line changes or output load changes is determined by the small internal junction capacitances within the transistor. It is normally safe to assume transistor reaction times of microseconds or quicker. Now consider the speed with which the unregulated input voltage fluctuates because of ripple—typically 120 Hz or a complete cycle every $\frac{1}{120}$ second [8.3 milliseconds (ms)]. Thus the regulator input ripple fluctuations are slow in comparison to its reaction time. It should now be clear that the regulator serves to reduce ripple and hence functions as a sophisticated "filter." A good regulator serves to reduce ripple by a factor of 100 or more.

EXAMPLE 3-4

You are to design a voltage regulator as indicated in Fig. 3-11. The load current varies between 0–1 A, and the unregulated dc input varies from 12–18 V for all line and load changes. The 8.5 V zener diode requires at least 1 mA of current to stay in its regulating region ($I_{ZK} = 1$ mA).

(a) Determine the value of R_S to ensure proper circuit operation.
(b) Determine the worst-case power dissipation of all circuit elements.

Solution:

(a) As previously stated, the worst-case conditions dictate that the value of R_S must be low enough to supply current to both the base of $Q1$ and the zener diode. This means that under worst-case conditions R_S must supply at least the $I_{ZK} = 1$ mA rating of $CR1$ plus the maximum base current:

$$I_{B\ max} = \frac{I_{L\ max}}{h'_{fe}} = \frac{1\ \text{A}}{50} = 20\ \text{mA}$$

$$I_{RS} = I_{ZK} + I_{B\ max} = 1\ \text{mA} + 20\ \text{mA} = 21\ \text{mA}$$

Now this 21 mA must be supplied by R_S under all conditions of input voltage variation—even when the input falls to 12 V, which causes the minimum voltage across R_S and hence the lowest value of current it will be able to supply. Therefore

$$R_S = \frac{12\ \text{V} - 8.5\ \text{V}}{21\ \text{mA}} = \frac{3.5\ \text{V}}{21\ \text{mA}} = 166\ \Omega$$

Should the value of R_S become any larger, the power supply would go out of regulation at $V_{IN} = 12$ V and $I_L = 1$ A, since the zener diode would not be supplied with

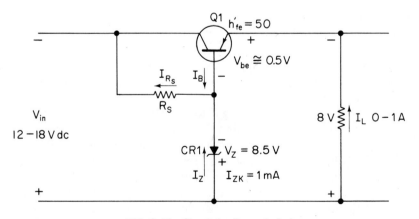

FIG. 3-11. Circuit for Example 3-4

the 1 mA it requires to regulate at the 8.5 V level. Therefore, R_S should be 166 Ω or slightly less to provide a margin of safety.

(b) The maximum power dissipation of R_S occurs when the voltage across it is maximum. Hence

$$P_{RS\ max} = \frac{(V_{RS\ max})^2}{R_S} = \frac{(V_{IN\ max} - V_Z)^2}{R_S}$$

$$= \frac{(18\ V - 8.5\ V)^2}{166\ \Omega} = 0.542\ W$$

Next consider the power dissipated in the zener diode. The voltage across it should remain at close to 8.5 V. The current varies from a low of 1 mA to a value not yet determined. The maximum value will occur when V_{IN} is maximum and the load current is minimum—in this example I_L goes down to zero, which means that I_E and hence I_B will be zero. If I_B is zero, all the current through R_S will pass through the diode:

$$I_{Z\ max} = I_{RS\ max} = \frac{V_{IN\ max} - V_Z}{R_S}$$

$$= \frac{18\ V - 8.5\ V}{166\ \Omega} = 57.2\ mA$$

$$P_{Z\ max} = V_Z I_{Z\ max} = 8.5\ V \times 57.2\ mA = 0.486\ W$$

The maximum power dissipation in $Q1$ will occur when its V_{CE} is maximum and when its emitter current is maximum. Its V_{CE} is equal to the input voltage minus V_L (Kirchhoff's voltage law):

$$V_{CE\ max} = V_{IN\ max} - V_L$$
$$= 18\ V - 8\ V = 10\ V$$

$$P_{Q1\ max} = V_{CE\ max} \times I_E = 10\ V \times 1\ A = 10\ W$$

3-5 Variable Output Regulated Power Supply with Overcurrent Protection

In this section a somewhat more sophisticated series regulated power supply is analyzed. The circuit illustrated in Fig. 3-12 is a regulated power supply with overcurrent protection and a variable output of 11–32 V. The circuit functions are identified within the dashed line blocks to correspond to the block diagram of Fig. 3-8. The circuit explanation will be done via these blocks in subsequent paragraphs.

1. Ac-to-dc converter. Transformer $T1$ steps down the ac line voltage to the appropriate level and the full-wave bridge diodes ($CR1$–$CR4$) provide the ac-to-dc conversion.

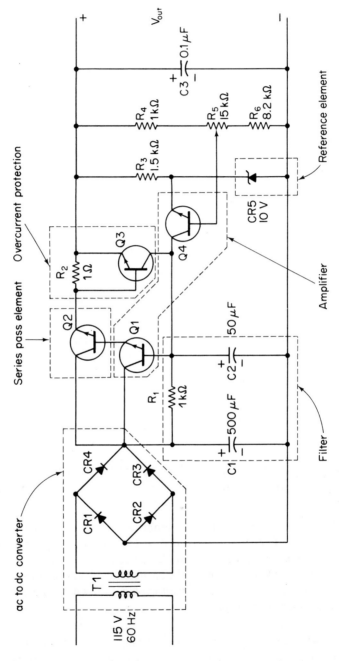

FIG. 3-12. Variable output regulated supply

2. Filter. Capacitors $C1$ and $C2$ filter the pulsating direct current to provide a relatively low level of ripple to the rest of the regulator. Keep in mind that the remaining circuitry (the regulator) serves to reduce this level of ripple so that the output voltage ripple is reduced further.

3. Reference element. Zener diode $CR5$ is the reference element against which the output is continually compared. The voltage from potentiometer $R5$'s wiper to ground is kept almost equal to the zener voltage, since the small base–emitter drop of $Q4$ is the only difference between these two voltages. Since the voltage from $R5$'s wiper is held relatively constant, the output voltage remains relatively constant, because the $R4$, $R5$, and $R6$ combination forms a voltage divider of the output voltage at $R5$'s wiper. Adjustment of the output voltage is thereby accomplished by adjusting $R5$.

4. Amplifier. Transistors $Q4$ and $Q1$ effectively amplify the error signal between the output voltage (through the $R4$, $R5$ and $R6$ voltage divider) and $CR5$'s zener voltage. This error signal is applied to $Q4$'s base–emitter bias and is amplified in the form of a variable collector current for $Q4$. If, for example, the regulated output voltage increases due to either a sudden line voltage increase or a sudden decrease in load, the voltage at $R5$'s wiper would increase correspondingly by voltage divider action. Since the zener voltage tends to remain constant, the base–emitter bias of $Q4$ would increase and $Q4$'s collector current would thereby increase. Thus $Q4$ amplifies the error signal and causes less current to be available for $Q1$'s base. Therefore, $Q1$ is also amplifying the error signal with its output, $Q1$'s emitter current, controlling the current flow through the series pass element, $Q2$. Since $Q1$ and $Q2$ both receive less base current as a result of the sudden increase in output voltage, less load current ($Q2$'s emitter current) is provided and the output voltage drops to about its original value. The circuit also serves to compensate for reductions in the output voltage in a similar, but reverse, fashion. Changing the setting of a potentiometer, $R5$, will change the bias on $Q4$ and hence adjust the output voltage.

5. Series pass element. The series pass element, $Q2$, receives the amplifier error signal from $Q1$ and uses it to control the output voltage as explained in the previous paragraph. It also functions as the last part of the amplifier section, since the error signal current is amplified by its h'_{fe} from $Q1$'s base to emitter.

The overcurrent protection is provided by $R2$ and $Q3$. Since $R2$ is in series with the output, the voltage across it is proportional to the output current. This serves to increase the base–emitter junction bias of $Q3$ as the load current increases, which starts turning it on as a voltage of about 0.5 V is reached. As $Q3$ is turned on harder by increasing load current, the base drive to $Q1$ is removed, which turns off power regulator $Q2$, thus limiting the output current. Resistor $R2$ is chosen, in this case, to limit the maximum load current to approximately 700 mA (V_{BE} of $Q3 = 0.7$ V). The curves shown in Fig. 3-13 show the effectiveness of the current limiting, as well as providing information from which the load regulation of the power supply can be calculated.

However, better detail on load regulation is provided by Fig. 3-14. The power supply illustrated in Fig. 3-12 is only one of the countless forms of voltage regulators. It is representative, however, of the general features contained in all of them.

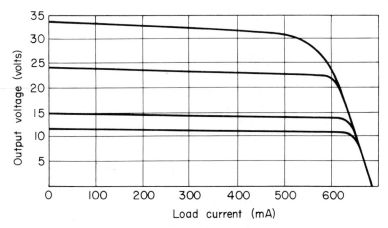

FIG. 3-13. Voltage output versus load current for regulator of Fig. 3-12 (Courtesy of Motorola Semiconductor Products, Inc.)

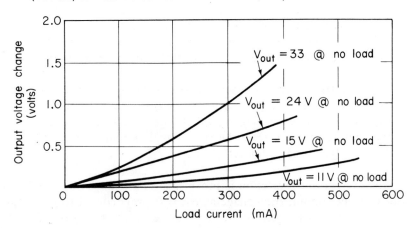

FIG. 3-14. Regulation versus load current for regulator of Fig. 3-12 (Courtesy of Motorola Semiconductor Products, Inc.)

EXAMPLE 3-5

Determine the load regulation of the power supply in Fig. 3-12 based on the data provided in Fig. 3-14 for a no-load voltage output of 33 and 11 V. Consider 400 mA as full load.

Solution:
When $V_L = 33$ V with no load,

$$\% \text{ load regulation} = \frac{\text{output voltage change} \times 100\%}{V_{FL}}$$
$$\simeq \frac{1.5 \text{ V} \times 100\%}{33 \text{ V} - 1.5 \text{ V}}$$
$$= 4.92\%$$

When $V_L = 11$ V with no load,

$$\% \text{ load regulation} = \frac{0.2 \text{ V}}{11 \text{ V} - 0.2 \text{ V}} \times 100\% = 1.86\%$$

3-6 Constant Current Supplies

Occasionally there is a demand for a source that delivers a constant current instead of a constant voltage. A common application of constant current supplies is in sweep circuits for cathode-ray tubes (CRT) in many pieces of electronic equipment. The constant current is used to charge a capacitor that drives the CRT, and this charging rate will be linear (hence a linear sweep) if the current supplied is truly constant.

Figure 3-15 shows an ideal current source. It has a source of current in parallel with an infinite source impedance, R_S. Recall from Section 3-2 that an ideal voltage source is a source of voltage in series with zero resistance. Since R_S is infinite in an ideal current source, all the source current will flow through the load, regardless of the load's value. The term ideal (perfect) current source then is used, since the load current is always constant. In practice, it is impossible to obtain an infinite source resistance current source,

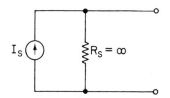

FIG. 3-15. Ideal current source

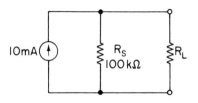

FIG. 3-16. Practical current source

just as it is a zero source resistance voltage source. Figure 3-16 shows a practical current source with a 100 kΩ source resistance. As long as the load R_L is relatively low, the load current remains relatively constant.

EXAMPLE 3-6

Determine the load current for the circuit shown in Fig. 3-16 for load values of 1, 5, 10, and 50 kΩ.

Solution:
The current divider rule can be used to determine the load current:

$$I_L = I_S \frac{R_S}{R_S + R_L}$$

For

$R_L = 1 \text{ k}\Omega, \quad I_L = 10 \text{ mA} \times \dfrac{100 \text{ k}\Omega}{100 \text{ k}\Omega + 1 \text{ k}\Omega} \simeq 9.99 \text{ mA}$

$R_L = 5 \text{ k}\Omega, \quad I_L = 10 \text{ mA} \times \dfrac{100 \text{ k}\Omega}{100 \text{ k}\Omega + 5 \text{ k}\Omega} \simeq 9.55 \text{ mA}$

$R_L = 10 \text{ k}\Omega, \quad I_L = 10 \text{ mA} \times \dfrac{100 \text{ k}\Omega}{100 \text{ k}\Omega + 10 \text{ k}\Omega} \simeq 9.1 \text{ mA}$

$R_L = 50 \text{ k}\Omega, \quad I_L = 10 \text{ mA} \times \dfrac{100 \text{ k}\Omega}{100 \text{ k}\Omega + 50 \text{ k}\Omega} \simeq 6.67 \text{ mA}$

The previous example shows that high values of source resistance are desirable for constant current supplies. Figure 3-17 shows a transistorized constant current supply in which the constant current is taken from the CB transistor's collector. Since the collector–base junction is reverse biased, this circuit exhibits the desired high value of source resistance.

In this circuit the voltage divider $R1$ and $R2$ sets the base at approximately

$$\frac{800 \text{ }\Omega}{800 \text{ }\Omega + 200 \text{ }\Omega} \times 20 \text{ V} = 16 \text{ V}$$

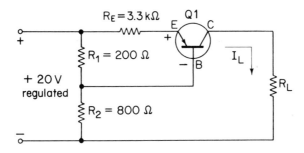

FIG. 3-17. Constant current supply

In using the voltage divider to determine the voltage at the base, we are assuming negligible loading effects on the divider by the base. This assumption is justified in this case. If we assume a 0.7 V drop from base to emitter, the voltage across R_E becomes 20 V − 0.7 V − 16 V = 3.3 V. Hence the emitter current is

$$I_E = \frac{3.3 \text{ V}}{3.3 \text{ k}\Omega} = 1 \text{ mA}$$

Note that we have calculated the emitter current without having made reference to any effect from R_L. Within certain upper bounds on the value of R_L, this circuit will deliver a constant current of 1 mA to the load. If, for instance, R_L got as high as 16 kΩ, then $V_L = 16$ kΩ × 1 mA = 16 V. With 3.3 V across R_E this means the voltage from collector to emitter would be 20 V − 3.3 V − 16 V = 0.7 V. The transistor is either saturated or very close to it at this point, and no further increase in R_L could be tolerated without having the current output fall off.

Notice that a voltage regulated supply is required to make the current regulator function properly. The voltage regulator need only correct for line input variations, since its load is constant as long as the current regulator is functioning properly.

EXAMPLE 3-7

Determine the minimum value which R_L could assume that would still allow proper circuit operation for the current regulator in Fig. 3-17.

Solution:
Examine the effects of using a load of zero resistance (i.e., a short circuit). In this case $V_L = 0$ and hence

$$V_{CE} = 20 \text{ V} - 3.3 \text{ V} = 16.7 \text{ V}$$

The emitter current is still 1 mA; hence the collector current flowing through the short is 1 mA and proper circuit operation is maintained, even down to a short-circuit load.

PROBLEMS

1. An 11 V battery has an internal impedance of 0.5 Ω. Determine the voltage it would deliver to loads of 1, 2, 5, 25, and 100 Ω.
2. Calculate the load regulation for the battery used in Problem 1 when its rating is 0–1 A. Express your answer as a percentage.
3. Calculate an appropriate value for R_S in the circuit shown in Fig. P3-3.

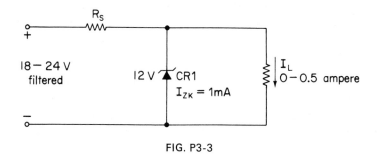

FIG. P3-3

4. Calculate the power rating required of R_S and $CR1$ in Problem 3.
5. Calculate the percentage regulation for Problem 4, given that the zener diode has an internal impedance Z_z of 0.5 Ω.
6. Calculate an appropriate value for R_S in the circuit shown in Fig. P3-6.

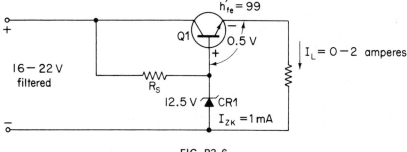

FIG. P3-6

7. Calculate the required power ratings for all circuit elements in Problem 6.
8. Repeat Problem 6 with the transistor's h_{fe} reduced from 99 to 49.
9. Explain why an electronic regulator serves to reduce the ripple level considerably.

10. Determine the current output of the constant current regulator shown in Fig. P3-10.

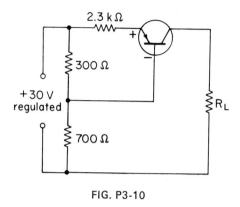

FIG. P3-10

11. Through what range of R_L will the circuit shown in Fig. P3-10 be effective?

4

Basic Transistor Amplifiers

4-1 Introduction

In this chapter an introduction to the three basic forms of junction transistor amplifiers is presented. Equations for their input impedance, voltage gain, and current gain are developed and applied in practical examples. The need for feedback for the common-emitter (CE) amplifier is illustrated, but the use of that feedback is not introduced until the following chapter.

The importance of the student's understanding of this chapter's concepts is extreme. It is the building block for future work and hence must be understood to allow for satisfactory future advancement.

4-2 Common-Base Amplifiers—DC Relations

Before a transistor can amplify an ac signal, it must be properly biased. That is, the proper dc relationships must be set up. Recall from Chapter 1 that this means the base–emitter junction must be forward biased and the base–collector junction must be reverse biased. Figure 4-1 shows a method that can be used to properly bias a CB amplifier. The 10 V battery in the

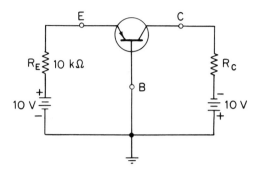

FIG. 4-1. Biasing a CB amplifier

emitter circuit will forward bias the base–emitter junction. Since the forward voltage drop of a silicon *pn* junction is about 0.7 V, the dc emitter current, I_E, can be calculated as

$$I_E = \frac{10 \text{ V} - 0.7 \text{ V}}{10 \text{ k}\Omega} = 0.93 \text{ mA}$$

Recall from Chapter 1 that $I_C = h_{FB}I_E$ and that h_{FB} is very nearly 1. Hence the collector current can be approximated with the value 0.93 mA:

$$I_C = h_{FB}I_E \simeq I_E$$

It should now be possible to calculate a value of R_C to provide a good Q-point voltage at the transistor's collector.

The ideal Q-point voltage is approximately one half the collector supply voltage or, in this case, -5 V. Since the ac output of the amplifier is at the collector for a CB circuit, this then allows for the greatest possible ac output signal before clipping will occur. For the case of a 10 V collector battery, an ac output signal of nearly 10 V p-p is possible before the transistor would be fully on (saturated) or fully off (cut off). If the transistor were saturated or cut off during either the highest positive or negative excursion of the ac output signal, clipping would occur.

To calculate the value of R_C that provides a collector Q-point of -5 V, divide the current through it into the voltage across it:

$$R_C = \frac{V_{RC}}{I_C} \simeq \frac{5 \text{ V}}{0.93 \text{ mA}} = 5.4 \text{ k}\Omega$$

Using a 5.4 kΩ resistor for R_C then results in an ideal bias of -5 V at the collector.

EXAMPLE 4-1

For the circuit shown in Fig. 4-2, calculate the dc voltage at the collector.

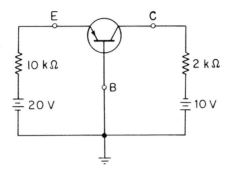

FIG. 4-2. Example 4-1

Solution:
The emitter current will be

$$I_E = \frac{20 \text{ V} - 0.7 \text{ V}}{10 \text{ k}\Omega} = 19.3 \text{ mA}$$

This is about equal to the collector current, and the voltage across R_C can be calculated as

$$V_{RC} = I_C R_C \simeq 19.3 \text{ mA} \times 2 \text{ k}\Omega$$
$$= 3.86 \text{ V}$$

Thus the voltage at the collector will be

$$V_C = V_{CC} - V_{RC}$$
$$= -10 \text{ V} + 3.86 \text{ V} = -6.14 \text{ V}$$

4-3 Equivalent Circuit for the CB Transistor Amplifier

In a CB amplifier the input is applied to the emitter lead, the base lead is common to input and output, and the output is taken at the collector, as shown in Fig. 4-3. This figure is a simplification, since no batteries or resistors for biasing are shown. However, since the input is the forward-biased base–emitter junction, it stands to reason that the input impedance will be largely determined by the low resistance of such a junction. The collector

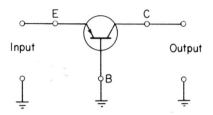

FIG. 4-3. Representation of a CB amplifier

circuit is a reverse-biased junction and hence has a very high output resistance. Since the collector has a current flow of h_{fb} times the input emitter current, it seems reasonable to represent the collector circuit by a current source. Figure 4-4 is an equivalent circuit for the CB amplifier and is some-

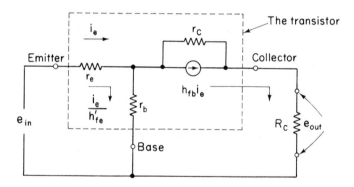

FIG. 4-4. T-equivalent circuit for CB configuration

times referred to as the T-equivalent circuit for the CB configuration. Notice the relations provided in Chapter 1 relating the base, emitter, and collector currents are used in this figure. The base current is equal to the emitter current divided by h_{fe}, and the collector current equals h_{fb} times the emitter current.

Now that we have introduced an equivalent circuit, we shall use it to develop some important relationships for the CB circuit. Applying Kirchhoff's voltage law to the left-hand loop of Fig. 4-4 yields

$$e_{in} = i_e r_e + \frac{i_e}{h'_{fe}} r_b \qquad (4\text{-}1)$$

Since e_{in}/i_{in} equals the input impedance, and i_{in} is the emitter current, i_e, we can divide both sides of Eq. (4-1) by i_e to solve for R_{in}:

$$R_{in} = r_e + \frac{r_b}{h'_{fe}} \qquad (4\text{-}2)$$

This input resistance is often provided in transistor data sheets as one of the hybrid (h) parameters, h_{ib}. Therefore, h_{ib} is equal to the R_{in} in Eq. (4-2) and is thus equal to the emitter resistance, r_e, plus the base resistance, r_b, divided by h_{fe}. Shockley, one of the discoverers of transistor action, has shown that the common-base ac input resistance, h_{ib}, is given by the following approximation:

$$h_{ib} \simeq \frac{0.026}{I_E} \qquad (4\text{-}3)$$

Note that I_E in Eq. (4-3) is in uppercase letters and thus refers to the dc or quiescent emitter current. This relationship will be of great usefulness to us in much of our future work and is therefore worthy of memorization. Note the similarity between Eqs. (4-3) and (1-1) developed for a diode.

We are now in a position to determine the various gains of the CB amplifier. By reference to Fig. 4-4 the current gain is simply equal to h_{fb}. Recall from Chapter 1 that $h_{fb} = h_{fe}/(h_{fe} + 1)$, and is therefore always slightly less than 1. The voltage gain is obtained by utilizing the TGIR [Eq. (1-19)]:

$$A_v = A_i \frac{R_L}{R_i} \qquad (1\text{-}19)$$

Substituting for A_i and R_i yields

$$A_v = h_{fb} \frac{R_L}{h_{ib}} \qquad (4\text{-}4)$$

Substituting Shockley's relation for h_{ib} ($h_{ib} = 0.026/I_E$) and approximating h_{fb} with 1, we have

$$A_v \simeq \frac{I_E R_C}{0.026} \simeq \frac{R_C}{h_{ib}} \qquad (4\text{-}5)$$

since $h_{ib} \cong 0.026/I_E$. Since we have a current gain of nearly 1, A_p is approximately equal to A_v ($A_p \cong A_v$). It is then evident that A_v and A_p are both dependent upon the value of R_C and the dc emitter current.

4-4 Common-Base Amplifiers

Figure 4-5 illustrates a CB amplifier that could be of practical value. The coupling capacitors $C1$ and $C2$ should be considered as short circuits to the ac signal and open circuits to the dc signal. The two batteries, along with R_E and R_C, serve to set up the required dc operating point for the amplifier. As explained in Section 4-2, it is normal to set up the operating point such that the transistor output terminal (in this case the collector) is at approximately

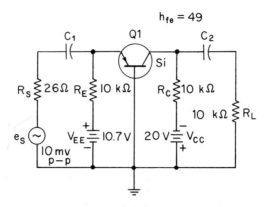

FIG. 4-5. Practical CB amplifier

one half the supply voltage or $V_{CC}/2$. This allows for the maximum possible output voltage swing before saturation or cutoff occurs in an amplifier. The results of allowing the transistor to become saturated or cut off would be to clip the output signal in its maximum positive and/or negative excursions. If the output signal is small (in the millivolt region), this biasing rule of thumb is not of much consequence. R_L is the ac load to the circuit and the ac signal is represented by e_{in} and its internal source resistance, R_S.

For the circuit of Fig. 4-5 we are given that

$$e_S = 10 \text{ mV p-p}$$
$$R_S = 26 \text{ }\Omega$$
$$R_E = 10 \text{ k}\Omega$$
$$R_C = 10 \text{ k}\Omega$$
$$R_L = 10 \text{ k}\Omega$$
$$h_{fe} = 49$$
$$V_{EE} = 10.7 \text{ V}$$
$$V_{CC} = 20 \text{ V}$$

We would like to determine the ac output voltage across R_L. The first step is to solve for the dc emitter current, from which we can solve for h_{ib}, the input resistance. The dc emitter current can be calculated as $(V_{EE} - V_{EB})/R_E$, since $V_{EE} - V_{EB}$ is the voltage drop across R_E. We shall assume a value for V_{EB} of 0.7 V, which is typical for the voltage across the forward-biased silicon pn junction:

$$I_E = \frac{10.7 \text{ V} - 0.7 \text{ V}}{10 \text{ k}\Omega} = 1 \text{ mA}$$

Now, by Shockley's relation we find $h_{ib} = 0.026/0.001$ A $= 26\,\Omega$. This is the input resistance to the transistor; thus the input impedance to the amplifier, including the bias resistor, R_E, is $26\,\Omega \| R_E$ with the battery appearing as a short circuit to the ac input signal. Since $26\,\Omega \| 10\,\text{k}\Omega \simeq 26\,\Omega$, the value of R_{IN} is simply h_{ib}, or $26\,\Omega$ in this case, and by voltage divider action we see that only half our source voltage, e_S, gets applied to the amplifier input because of the voltage split between the $26\,\Omega$ source resistance and the very low input resistance of the CB amplifier of $26\,\Omega$. Thus the input voltage to the transistor's emitter is only 5 mV p-p. Refer to Fig. 4-6 as an aid in understanding

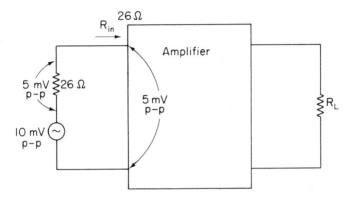

FIG. 4-6. Voltage divider effect at input of amplifier

this effect. This low input impedance is a characteristic of CB amplifiers that limits their usefulness, since it takes such a low impedance to effectively drive them.

The voltage gain of the transistor alone, A_v, is

$$A_v = \frac{I_E R_C}{0.026} \simeq 0.001 \times \frac{10\,\text{k}\Omega}{0.026}$$
$$= 384$$

The total voltage gain of the amplifier, G_v, must be calculated, including the effects of the external ac load, R_L, and the fact that only one half of e_S actually gets applied to the transistor:

$$G_v = \left(\frac{26\,\Omega}{26\,\Omega + 26\,\Omega}\right)\frac{I_E(R_C \| R_L)}{0.026} = \frac{1}{2}\frac{0.001\,\text{A} \times 5\,\text{k}\Omega}{0.026}$$
$$= 96$$

Hence the output voltage is 10 mV p-p $\times$ 96 $=$ 960 mV p-p, or 0.96 V p-p. A quick check of our collector bias point shows it to be ideally set up to a

70 Basic Transistor Amplifiers

-10 V level, which is half the supply voltage, or

$$V_C = V_{CC} - V_{RC} \simeq -20 \text{ V} - (-1 \text{ mA} \times 10 \text{ k}\Omega) = -10 \text{ V}$$

One last noteworthy fact which we can learn from this amplifier is that the gain of a CB amplifier is not very dependent on the transistor's h_{fe}. Since h_{fe} varies widely with temperature and from one transistor to another, this gain stability is a very good attribute of the CB amplifier. Notice that in no place did we use the h_{fe} of the transistor to make a calculation.

EXAMPLE 4-2

(a) For the circuit of Fig. 4-7, solve for the transistor's voltage gain, A_v.
(b) Solve for the circuit's voltage gain, G_v.
(c) Sketch the input and output voltage waveform and show the dc levels.

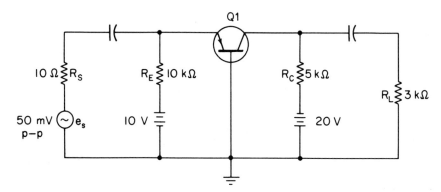

FIG. 4-7. Circuit for Example 4-2

Solution:
(a) First solve for I_E:

$$I_E = \frac{V_{R_E}}{R_E} = \frac{10 \text{ V} - 0.7 \text{ V}}{10 \text{ k}\Omega} = 0.93 \text{ mA}$$

$$A_v \simeq \frac{I_E R_C}{0.026} = \frac{0.93 \text{ mA} \times 5 \text{ k}\Omega}{0.026} \quad (4\text{-}5)$$

$$= 179$$

(b) $\quad G_v = A_v \dfrac{R_{IN}}{R_{IN} + R_S} \quad$ by voltage divider action on e_s

$$R_{IN} = h_{ib} = \frac{0.026}{I_E} = \frac{0.026}{0.93 \text{ mA}}$$

$$= 28 \Omega$$

$$G_v = \frac{28\ \Omega}{10\ \Omega + 28\ \Omega} \times \frac{0.93\ \text{mA} \times (5\ \text{k}\Omega \parallel 3\ \text{k}\Omega)}{0.026}$$

$$= 50$$

Note also that the load on the overall amplifier is $5\ \text{k}\Omega \parallel 3\ \text{k}\Omega$ and not just the $5\ \text{k}\Omega$ of part (a).

(c)
$$e_{\text{out}} = e_{in}G_v$$
$$= 50\ \text{mV p-p} \times 50$$
$$= 2.5\ \text{V p-p}$$

The input and output waveforms will be in phase. As i_e increases, so will i_c; therefore, an increase in input voltage results in an increase in output voltage. The input and output voltage waveforms are shown in Fig. 4-8. A last check should be made to

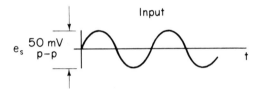

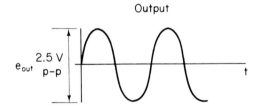

FIG. 4-8. Input and output waveforms for Example 4-2

ensure that the bias point is such as to allow a 2.5 V p-p swing without clipping:

$$I_C \simeq I_E = 0.93\ \text{mA}$$

The voltage at the collector will be $[-20\ \text{V} - (-I_C R_C)]$, which is $[-20\ \text{V} - (-4.65\ \text{V})] = -15.35\ \text{V}$. Therefore, the composite ac and dc signal at the collector of $Q1$ will be as shown in Fig. 4-9. As can be seen, no clipping has occurred. The signal across R_L is the 2.5 V p-p signal centered at ground, since the capacitor blocks the -15.35 V dc level. This is shown in Fig. 4-9.

72 Basic Transistor Amplifiers

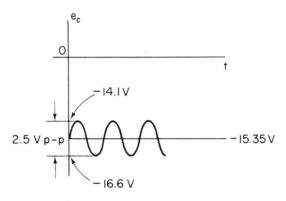

FIG. 4-9. Composite signal at collector of Q1 for Example 4-2

4-5 Common-Collector Configuration

The common-collector (CC) amplifier is often referred to as the emitter follower. It is so termed because the output voltage at the emitter "follows" the input voltage at the base. This circuit is of great importance in electronics, even though it offers no voltage gain. The reasons for this will become apparent as it is applied throughout the remainder of this text. Figure 4-10a shows a simple common-collector amplifier with the dc biasing resistors R_A and R_B, and part b shows an ac equivalent circuit for a transistor in a CC con-

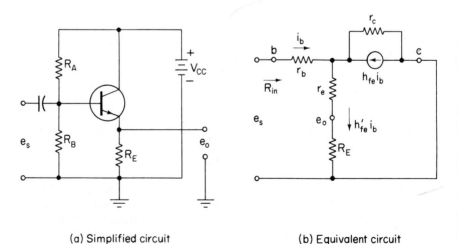

(a) Simplified circuit   (b) Equivalent circuit

FIG. 4-10. Common-collector amplifier

figuration along with an emitter resistor. In much the same fashion as we did for the CB circuit, we shall develop an expression for the input impedance of the CC circuit.

Looking at the input loop of Fig. 4-10a, by Kirchhoff's voltage law,

$$e_{in} = i_b r_b + h'_{fe} i_b (r_e + R_E)$$

Since $R_{in} = e_{in}/i_b$, we have

$$R_{in} = r_b + h'_{fe}(r_e + R_E) \tag{4-6}$$

Now recall from our work with the CB circuit that $h_{ib} = r_e + (r_b/h'_{fe})$. If we multiply both sides of this equation through by h'_{fe}, we have

$$h'_{fe} h_{ib} \simeq h'_{fe} r_e + r_b \tag{4-7}$$

Combining Eqs. (4-6) and (4-7), we have

$$R_{in} = h'_{fe}(h_{ib} + R_E) \tag{4-8}$$

This is an interesting result that bears a little analysis. It tells us that the CB input impedance, h_{ib}, can be utilized to determine the input impedance of the CC circuit. This is a very practical result, because h_{ib} is such an easy parameter to determine. Once you know the dc emitter current I_E, h_{ib} follows from Shockley's relation:

$$h_{ib} = \frac{0.026}{I_E} \tag{4-3}$$

In some instances h_{ib} is very small with respect to R_E so that Eq. (4-8) reduces to

$$R_{IN} \simeq h'_{fe} R_E \tag{4-9}$$

The current gain of the CC circuit simply equals h'_{fe}, as can be determined from the equivalent circuit:

$$A_i = h'_{fe} \tag{4-10}$$

Substituting into the TGIR, we can now solve for A_v as

$$A_v = A_i \frac{R_E}{R_{in}} = \frac{h'_{fe} R_E}{h'_{fe}(h_{ib} + R_E)}$$
$$= \frac{R_E}{h_{ib} + R_E} \tag{4-11}$$

This shows us that the voltage gain is always less than 1, but since h_{ib} is usually small with respect to R_E, it is normally only slightly less than 1. Therefore, in most of our calculations we shall assume a voltage gain for the CC circuit of 1. Since the output voltage (emitter voltage) is nearly the same as the input voltage (base voltage), we can say that the emitter follows the base, and hence the terminology emitter follower.

The output impedance of the CC circuit is R_E in parallel with the impedance looking into the emitter of the transistor. Looking into the transistor's emitter lead, h_{ib} plus the impedance of the external base resistances divided by h'_{fe} is seen. If R_{base} represents these base impedances,

$$R_{\text{out}} = R_E \left\| \left(h_{ib} + \frac{R_{\text{base}}}{h'_{fe}} \right) \right. \quad (4\text{-}12)$$

4-6 Common-Collector Applications

A question that might logically be asked at this point is, of what value is a circuit that has no voltage gain? The answer is that the circuit has the ability to transform from a high impedance level down to an impedance level a factor of h'_{fe} less.

The circuit in Fig. 4-11 shows a practical CC amplifier with an ac load R_L connected through a coupling capacitor, C_c. Resistors R_A and R_B are

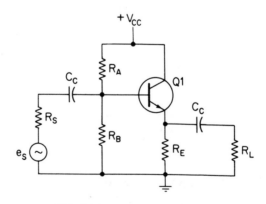

FIG. 4-11. Practical CC amplifier

the biasing resistors, and this form of biasing is called voltage divider biasing. Notice that the collector is common to the input and output only in the sense that the impedance of the battery V_{CC} is very close to zero. Thus the name common collector requires a small amount of imagination to justify. The coupling capacitor isolates the driving source (in this case e_s) from the rest of the circuit with respect to direct current, but we assume it to be a short circuit to our ac signal.

The generally most desirable dc operating point for the output terminal is, as in the CB amplifier, about one half the supply voltage. This allows for the maximum possible output swing without clipping and with minimum distortion.

EXAMPLE 4-3

A 1 kΩ load is to be driven from a 5 V p-p source that has an R_s of 2.5 kΩ.

(a) What output voltage would result from a direct drive?

(b) Design a CC amplifier to effect an impedance transformation to keep the output voltage from loading down, as it did in part (a) of this example.

(c) Calculate the output impedance of the circuit designed in part (b).

Solution:

(a) Referring to Fig. 4-12, we see that with direct drive

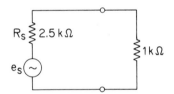

FIG. 4-12. Example 4-3a

$$e_{out} = e_s \frac{1 \text{ k}\Omega}{2.5 \text{ k}\Omega + 1 \text{ k}\Omega}$$
$$= 5 \text{ V p-p} \frac{1 \text{ k}\Omega}{3.5 \text{ k}\Omega}$$
$$= 1.43 \text{ V p-p}$$

This is generally an unacceptable result because of the severe loading effect on the driving impedance which results in a small output voltage.

(b) Figure 4-13 shows the general form our design will take. Note that we are not allowing direct current to flow in the load resistor R_L. Assume that $Q1$ has an h_{fe} of 199, and the biasing should result in a dc level at the emitter of 6 V.

$$R_{in} = h'_{fe}[h_{ib} + (R_E \| R_L)]$$

$$h_{ib} = \frac{0.026}{I_E} = \frac{0.026}{6 \text{ V}/1 \text{ k}\Omega} = 4.34 \text{ }\Omega \tag{4-8}$$

Obviously, h_{ib} is so much smaller than $R_E \| R_L$ that we are justified in using the approximate form for the CC input impedance:

$$R_{in} \simeq h'_{fe}(R_E \| R_L) \tag{4-9}$$
$$= 200 \times 500 \text{ }\Omega = 100 \text{ k}\Omega$$

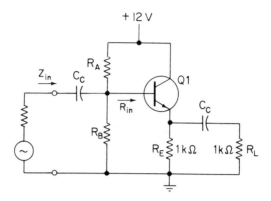

FIG. 4-13. Example 4-3b

The overall input impedance to the amplifier, Z_{in}, is equal to the parallel combination of R_A, R_B, and R_{in}. To minimize loading effects on the high impedance source, Z_{in} should be an order of magnitude higher than R_S, so our design goal will be a Z_{in} of 25 kΩ or greater. Let $R_B = 120$ kΩ and solve for the value of R_A that provides the desired 6 V bias at the emitter. Assuming a silicon transistor, a 6 V level at the emitter would mean about 6.7 V at the base. By voltage divider action,

$$V_B = V_{CC} \frac{R_B \| R_{IN}}{(R_B \| R_{IN}) + R_A}$$

$$6.7\text{ V} = \frac{12\text{ V}\left(\frac{120\text{ k}\Omega \times 100\text{ k}\Omega}{120\text{ k}\Omega + 100\text{ k}\Omega}\right)}{\left(\frac{120\text{ k}\Omega \times 100\text{ k}\Omega}{120\text{ k}\Omega + 100\text{ k}\Omega}\right) + R_A}$$

$$= \frac{12\text{ V} \times 54.5\text{ k}\Omega}{54.5\text{ k}\Omega + R_A}$$

$$R_A = 43\text{ k}\Omega$$

Therefore,

$$R_{in} = R_A \| R_B \| R_{in}$$
$$= 43\text{ k}\Omega \| 120\text{ k}\Omega \| 100\text{ k}\Omega$$
$$= 24\text{ k}\Omega$$

and assuming that $A_v = 1$, G_v will equal

$$\frac{Z_{in}}{Z_{in} + R_S} = \frac{24\text{ k}\Omega}{24\text{ k}\Omega + 2.5\text{ k}\Omega} \simeq 0.95$$

Therefore,

$$e_{out} = G_v e_s$$
$$= 0.95 \times 5\text{ V p-p}$$
$$= 4.75\text{ V p-p}$$

(c) The output impedance will be

$$R_{out} = R_E \| \left(h_{ib} + \frac{R_{base}}{h'_{fe}} \right)$$

$$= 1\text{ k}\Omega \| \left(4.34\ \Omega + \frac{43\text{ k}\Omega \| 120\text{ k}\Omega \| 2.5\text{ k}\Omega}{200} \right) \quad (4\text{-}12)$$

$$= 1\text{ k}\Omega \| (4.34\ \Omega + 11.9\ \Omega)$$

$$\simeq 16.2\ \Omega$$

Example 4-3 illustrates the ability of the emitter follower to transform from a high impedance to a low impedance with very little loss of signal. In many cases the driving source is another stage of an amplifier system, as shown in Fig. 4-14. This can often simplify the design of an emitter follower. If, for example, the output of the driving stage were biased at about one half the supply voltage, it might be possible to drive directly into the emitter follower without the need for the biasing resistors. An analysis of this circuit will show that an improvement in performance also occurs.

If the collector of $Q1$ in Fig. 4-14 is at 6 V, then the base of $Q2$ is also at 6 V, and we would expect the emitter of $Q2$ to be at about 5.3 V. This is an

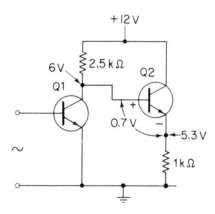

FIG. 4-14. Common-emitter amplifier stage driving directly into an emitter follower

acceptable bias point, being close to the midpoint of the supply voltage. The loading effects of the emitter follower on $Q1$ are almost negligible now, since instead of an input impedance of 24 kΩ, as with the biasing resistors in Example 4-3, the input impedance is simply $h'_{fe} \times 1$ kΩ or 200 kΩ if $h'_{fe} = 200$. Not only have we been able to simplify the circuit by eliminating the need for two biasing resistors and a coupling capacitor, but an improvement in performance has been made because of the increased input impedance.

Determination of the current gain for Example 4-3 reveals several important ideas. The current gain of the transistor (A_i) is equal to h'_{fe}, as indicated by Eq. (4-10). However, the overall current gain of this amplifier is something else again. You must remember that we are dealing with more than just a transistor, as the biasing resistors will also draw some of the current from the source. From the TGIR we know that

$$G_i = G_v \frac{R_{in}}{R_L}$$

The overall input impedance for Example 4-3 is 24 kΩ, and G_v was 0.95. Therefore,

$$G_i = 0.95 \times \frac{24 \text{ k}\Omega}{1 \text{ k}\Omega} = 23$$

This is obviously less than h'_{fe}. Figure 4-15 should help you see why the overall current gain is less than h'_{fe}. Notice that the battery has been replaced by

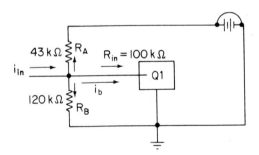

FIG. 4-15. Simplification of Fig. 4-11 for the purpose of visualizing G_i

a short circuit (its approximate impedance) and $Q1$ by its R_{in}. Now by the current divider rule, the current actually entering the base lead that will be amplified by the h'_{fe} factor will be

$$i_{base} = i_{in} \frac{R_A \| R_B}{(R_A \| R_B) + R_{in}}$$
$$= i_{in} \frac{31.7 \text{ k}\Omega}{31.7 \text{ k}\Omega + 100 \text{ k}\Omega} = 0.24 \, i_{in}$$

Therefore, the overall current gain of the complete circuit is

$$G'_i = 0.24 \times h'_{fe} = 0.24 \times 200 \times \tfrac{1}{2} = 24$$

The $\tfrac{1}{2}$ multiplying factor is the result of the even split of emitter current between R_E(1 kΩ) and the load R_L(1 kΩ).

The result is in close agreement with the TGIR solution, as it should be. Obviously, solving for G_i was a lot easier with the TGIR than by employing the current divider, and this gives you an idea of the TGIR's usefulness.

4-7 Common-Emitter Amplifiers—Introduction

The third and final version of junction transistor amplifiers is the common emitter (CE). It is by far the most widely used configuration. It provides both voltage and current gain, as compared to the other two forms which provide gains in voltage *or* current, but not both. Since power gain is equal to the product of voltage gain and current gain, the CE circuit provides a large amount of power gain as compared to the CB or CC circuits. In the CE circuit the emitter is common to both input and output, and the output is taken at the collector. In fact, a good rule of thumb for differentiating between CC and CE versions is to note where the output is taken; if the output is at the emitter, it is a CC circuit, and if the output is at the collector, it is a CE circuit. This is the best way to identify a circuit type, since determination of the common lead is often obscured by circuit peculiarities.

To determine an expression for input resistance, we shall once again resort to the equivalent-circuit technique. Figure 4-16a shows a CE amplifier minus necessary biasing circuitry, and Fig. 4-16b provides the equivalent

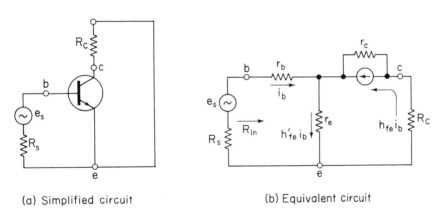

(a) Simplified circuit (b) Equivalent circuit

FIG. 4-16. Common-emitter amplifier

circuit for this amplifier. Taking the sum of voltage drops in the input side of the equivalent circuit,

$$e_{in} = i_b r_b + h'_{fe} i_b r_e \qquad (4\text{-}13)$$

Since $R_{in} = e_{in}/i_{in}$ and $i_{in} = i_b$,

$$R_{in} = \frac{e_{in}}{i_b} = r_b + h'_{fe}r_e \tag{4-14}$$

Since $h_{ib} = r_e + r_b/h'_{fe}$, we can manipulate Eq. (4-14) into

$$R_{in} = h'_{fe}h_{ib} \tag{4-15}$$

Once again we find that an easily determined parameter of the CB circuit is useful in calculations with another configuration. The extreme usefulness of this fact will make itself apparent over the next few chapters.

The device current gain of the CE configuration, A_i, is simply h_{fe}, since $i_c = h_{fe}i_b$ and i_c is the ac output current and i_b is the input current. Using the TGIR, we can develop an expression for the voltage gain:

$$A_v = h_{fe}\frac{R_L}{R_{in}} \tag{4-16}$$

Since $R_{in} = h'_{fe}h_{ib}$, we can say that

$$A_v \simeq \frac{h_{fe}R_C}{h'_{fe}h_{ib}} \tag{4-17}$$

But $h_{fe}/h'_{fe} \cong 1$, so then

$$A_v \simeq \frac{R_C}{h_{ib}} \tag{4-18}$$

Note that this is only the device gain without considering external loads or loading effects on the signal source.

The output impedance (R_o) of the CE amplifier is the impedance seen looking into the output lead, which is R_C in parallel with the reverse biased collector junction (a very high impedance). Therefore,

$$R_o \cong R_C \tag{4-19}$$

4-8 Common-Emitter Calculations

Figure 4-17 shows typical *npn* and *pnp* CE amplifiers with the necessary biasing circuitry. The only difference between the two circuits is the transistor polarity and the subsequent required change in battery polarity to provide the necessary forward bias to the base–emitter junction and reverse bias to the collector–base junction. Notice that a separate ac load is connected from

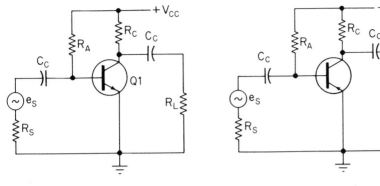

(a) Using npn transistor (b) Using pnp transistor

FIG. 4-17. Common-emitter amplifier

the collector, through a coupling capacitor, to ground. The ac collector impedance is now equal to $R_C \| R_L$, and hence the stage voltage gain is

$$G_v = \frac{R_C \| R_L}{h_{ib}} \qquad (4\text{-}20)$$

EXAMPLE 4-4

Given the following circuit values, determine the output voltage for the amplifier of Fig. 4-17a.

$e_s = 10 \text{ mV p-p}, \quad R_C = 10 \text{ k}\Omega, \quad R_L = 10 \text{ k}\Omega, \quad R_S = 1 \text{ k}\Omega,$

$h_{fe} = 100, \quad V_C = 6 \text{ V}, \quad V_{CC} = 12 \text{ V}$

Solution:

$$I_C = \frac{V_{CC} - V_C}{R_C} = \frac{6 \text{ V}}{10 \text{ k}\Omega} = 0.6 \text{ mA}$$

$$h_{ib} = \frac{0.026}{0.6 \times 10^{-3}} = 43.3 \text{ }\Omega$$

$$G_v = \frac{10 \text{ k}\Omega \| 10 \text{ k}\Omega}{43.3 \text{ }\Omega} \qquad (4\text{-}21)$$

$$= 115$$

This is the gain of the signal that gets applied to the transistor. Since the amplifier has some finite value of input impedance, and the signal source has an internal impedance, R_S, we know a portion of e_s will be dropped across R_S. Figure 4-18 provides a visualization of this effect.

$$e_{in} = e_s \frac{Z_{in}}{R_S + Z_{in}}$$

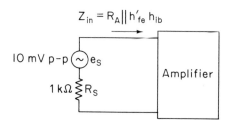

FIG. 4-18. Example 4-4

Z_{in} equals the parallel combination of the bias resistor R_A and the impedance seen looking into the base of $Q1$. We do not know the value of R_A, but since we know the dc collector voltage, we can work back and determine the value of R_A to provide this bias condition:

$$I_B = \frac{I_C}{h_{fe}}$$

$$= \frac{0.6 \text{ mA}}{100} = 0.006 \text{ mA}$$

$$R_A = \frac{V_{CC} - V_{BE}}{I_B} = \frac{12 \text{ V} - 0.7 \text{ V}}{0.006 \text{ mA}}$$

$$= 1.89 \text{ M}\Omega$$

Therefore,

$$Z_{in} = 1.89 \text{ M}\Omega \| h'_{fe} h_{ib} = 1.89 \text{ M}\Omega \| 101 \times 43.3 \text{ }\Omega$$
$$= 4.37 \text{ k}\Omega$$

Then, by voltage divider action,

$$e_{in} = 10 \text{ mV p-p} \times \frac{4.37 \text{ k}\Omega}{4.37 \text{ k}\Omega + 1 \text{ k}\Omega} = 8.17 \text{ mV}$$

Hence the overall voltage gain, G_{voa}, will be

$$G_{voa} = \frac{8.17}{10} \times 115 = 93.8$$

Therefore,

$$V_L = 10 \text{ mV p-p} \times 93.8 = 0.938 \text{ V p-p}$$

We can make an important conclusion from Example 4-4. The correct bias point ($V_C = 6$ V) was obtained with a bias resistor, R_A, of 1.89 MΩ.

However, if the dc collector current were to drift or if a replacement transistor were used having a different value of h_{fe}, the bias point would change appreciably. Consider a high volume production CE amplifier where the transistor called out has an h_{fe} range from 30–100. This is a typical h_{fe} spread and, for the circuit of our previous example, would necessitate hand-selected values of R_A for different transistors to provide the required 6 V bias point. This is not to mention the drift due to temperature variations after setup. It should also be mentioned that the amount of ac output voltage is also affected by the value of h_{fe}.

This extreme circuit performance dependence on h_{fe} in the CE circuit is overcome by the use of feedback. We shall discuss this in detail in the following chapter. It is interesting to note that the CB and CC circuits already discussed do not suffer to a great extent from this problem of bias stability and hence are practical circuits as presented in this chapter.

The CE amplifier has the characteristic of inverting the input signal. That is, the output signal is 180° out of phase with the input signal. When the input voltage to an *npn* CE amplifier goes positive, it increases the base–emitter forward bias and hence greater collector current flows. This increases the voltage drop across the collector resistor, which means that the collector voltage to ground (the output) decreases. Hence the output decreases as the input increases and vice versa. This characteristic is not true of the CB and CC configurations.

EXAMPLE 4-5

Determine the drift in bias point and the change in ac output voltage for Example 4-4 if h_{fe} were changed to 50.

Solution:
I_B is not changed by the change in h_{fe}:

$$I_B = \frac{V_{CC} - V_{BE}}{R_B} = 0.006 \text{ mA}$$

However, I_C will be halved since h_{fe} is cut in half:

$$I_C = h_{fe} I_B = 50(0.006 \text{ mA})$$
$$= 3 \text{ mA}$$
$$V_C = 12 \text{ V} - V_{RC}$$
$$= 12 \text{ V} - 10 \text{ k}\Omega \times 0.3 \text{ mA}$$
$$= 9 \text{ V}$$

Thus our Q-point has shifted from 6 to 9 V, which would cause serious clipping if the ac output was greater than 6 V p-p. Now to determine the change in ac output voltage, we first redetermine h_{ib}:

$$I_E \simeq I_C = 0.3 \text{ mA}$$

$$h_{ib} = \frac{0.026}{0.3 \text{ mA}} = 86.6$$

$$G_v = \frac{R_C \| R_L}{h_{ib}} = \frac{10 \text{ k}\Omega \| 10 \text{ k}\Omega}{86.6 \text{ }\Omega} = 57.5$$

$$Z_{in} = R_A \| h'_{fe} h_{ib} = 1.89 \text{ M}\Omega \| (51 \times 86.6 \text{ }\Omega)$$
$$= 4.4 \text{ k}\Omega$$

The overall voltage gain is

$$G_{voa} = \frac{Z_{in} G_v}{Z_{in} + R_S} = 57.5 \frac{4.4 \text{ k}\Omega}{4.4 \text{ k}\Omega + 1 \text{ k}\Omega} = 46.8$$

$$V_L = G_{voa} \times e_s$$
$$= 46.8 \times 10 \text{ mV p-p}$$
$$= 0.468 \text{ V p-p}$$

The approximate halving of the voltage gain was caused by the doubling of h_{ib} as a result of the halving of I_C. Since voltage gain is inversely proportional to h_{ib} in the non-feedback CE configuration, the gain was roughly cut in half.

EXAMPLE 4-6

(a) Determine the overall current gain, G_{ioa}, for Example 4-5.

(b) Sketch the input voltage and the composite voltage waveform at $Q1$'s collector on the same set of axes.

Solution:
(a) From the TGIR

$$G_{ioa} = G_{voa} \frac{Z_{in}}{R_L}$$

$$G_{ioa} = 57.5 \frac{4.4 \text{ k}\Omega}{5 \text{ k}\Omega} = 50$$

This is also the transistor's h_{fe} and is therefore an expected result, since virtually all the input current flows into the base lead, with the high value of R_A minimizing the only other alternative flow of input current.

(b)

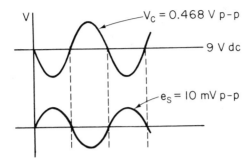

FIG. 4-19. Example 4-6

Notice the phase inversion between input and output voltage waveforms in Fig. 4-19. This is a characteristic of CE amplifiers that is not true of the CB and CC configurations.

PROBLEMS

1. Redraw the circuit of Fig. 4-5 to make it work using a *npn* transistor instead of the *pnp* variety shown.
2. Why does biasing an amplifier output at one half the supply voltage minimize clipping effects?
3. Why is it not *entirely* true to say that h_{fe} has no effect on the gains of a CB amplifier?
4. (a) In Example 4-2, determine the maximum level that e_s could be raised to without causing clipping in the output.
   (b) Would the amplifier go into saturation or cutoff first?
   (c) Provide three different design changes that could remedy this situation.
5. Design an emitter-follower circuit to transform from a 5 V p-p, 500 Ω source down to a 200 Ω load. The output voltage may not fall below 4.5 V p-p, the transistor has an h_{fe} of 99, and one 12 V battery is available for the job.
6. Calculate the current gain for Problem 5 using the TGIR and one other method.
7. Redraw your circuit from Problem 5 to make it operable with a transistor of opposite polarity from what you originally used.

8. For the amplifier shown in Fig. P4-8 determine the appropriate value for R_A, the ac output voltage, and the current gain.

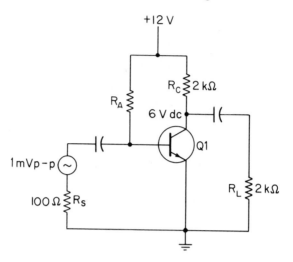

FIG. P4-8.

9. Determine the change in bias point and ac output voltage for Problem 8 if h_{fe} were cut in half.

10. Sketch the input and collector voltage waveform for Problem 9 on the same set of axes, and explain the reason for the 180° phase shift between the two.

11. Determine the effect on the ac output voltage for Problem 8 if
    (a) R_A increased.
    (b) h_{fe} increased.
    (c) V_{CC} increased.
    (d) R_L decreased.
    (e) R_C decreased.
    (f) R_S decreased.

5

Single-Stage Feedback Amplifiers

5-1 Introduction

The CE amplifier of Chapter 4 had two characteristics that kept it from being a practical, useful circuit. Those shortcomings were (1) an inability to operate over varied temperatures, and (2) the fact that it does not lend itself to mass production—a necessity to make a circuit economically feasible.

These problems both stem from the high variability of h_{fe}, which results in unpredictable gains and changes in Q-point. A transistor's h_{fe} may vary by a 2:1 range for a temperature increase of from room temperature to 125°C. Similarly, the h_{fe} of a specific type of transistor usually varies by a 2:1 factor from unit to unit. This last problem could be overcome by paying an exorbitant price for specially selected h_{fe} range transistors, but fortunately a better solution exists for these ills.

The solution is to simply make the CE transistor amplifier insensitive to transistor characteristics and instead, through the use of feedback, make the amplifier's performance dependent on bias and circuit resistor values. Resistor values are more stable and predictable than transistor characteristics. Two other side benefits of feedback are often an improved response to high and low frequencies (at the expense of gain) and an increased input impedance. *Feedback* occurs when a part of an output signal is fed back to the input signal, causing a change in the output signal. When this fed-back signal causes a reduction in the output signal, we say that *negative feedback* is taking place. It is this type of feedback that forms the basis for our work in this chapter.

Despite the rosy picture thus far painted for feedback, there is one price to pay for this cure-all. The price is simply a loss of gain. In fact, the effectiveness of an amplifier's improvement is roughly proportional to the loss in gain introduced. This sacrifice for improved performance can be easily compensated for by adding the necessary number of stages to make up for the loss in gain introduced by feedback.

In this chapter we introduce the feedback concept and apply it to the workhorse of electronic amplification, the single-stage CE bipolar transistor amplifier. The use of feedback as it applies to FETs and multistage amplifiers is left for subsequent chapters.

5-2 Feedback Principles

Figure 5-1 is a block diagram of a feedback amplifier. This amplifier has an *open-loop* gain (gain without feedback) of G_v. It feeds back a propor-

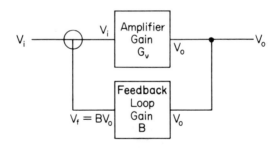

FIG. 5-1. Feedback amplifier

tion of the output to the input. This proportion B is called the *feedback factor*. In most texts it is given the symbol β, but to avoid confusion with the current gain of a CE amplifier it will be signified by the letter B. The feedback voltage V_f equals BV_o. Hence

$$B = \frac{V_f}{V_o} \tag{5-1}$$

so that the resultant input voltage V_i to the amplifier, referring to Fig. 5-1, is

$$V_i = V_g + V_f = V_g + BV_o \tag{5-2}$$

Since $V_o = G_v V_i$, we can write

$$G_v V_i = G_v(V_g + BV_o) \tag{5-3}$$

If G'_v is the gain with feedback (V_o/V_g), then Eq. (5-3) can be manipulated to give

$$G'_v = \frac{G_v}{1 - BG_v} \tag{5-4}$$

where G'_v is the gain with feedback V_o/V_g. The gain with feedback G'_v is often termed the *closed-loop* gain since it is the gain with the feedback loop "closed" around the amplifier. This compares to the open-loop gain G_v that would exist if the feedback loop were "open" or removed.

Equation (5-4) is a general equation and is applicable not only to negative feedback but also to the positive feedback that exists in oscillators. With positive feedback, the fed-back signal is in phase with the input, causing G'_v to be greater than G_v. In this chapter we are concerned with negative feedback where the fed-back signal is out of phase with the generator voltage, making V_i smaller than it otherwise would be without feedback. Recall that two signals 180° out of phase, but equal in frequency, will tend to cancel themselves with the resultant having the phase of the larger of the two signals. This resultant signal would have an amplitude equal to the difference of the two original signals. This phase difference means that the *feedback factor* B is negative, and hence the denominator of Eq. (5-4) will be greater than 1; therefore, G'_v is less than G_v.

An interesting condition exists when we make the term BG_v of Eq. (5-4) quite large with respect to 1. Since

$$G'_v = \frac{G_v}{1 - BG_v}$$

and if

$$BG_v \gg 1$$

then

$$G'_v \simeq \frac{G_v}{-BG_v}$$

and

$$G'_v = -\frac{1}{B} \tag{5-5}$$

We now have the condition in which the gain of the amplifier is completely dependent on the feedback factor B and completely independent of G_v! Since B is usually determined by resistor values, we have arrived at our goal of overall amplifier gain completely independent of the variable parameters of transistors.

EXAMPLE 5-1

An amplifier design has an open loop gain that can vary from 50 to 200 for variations in temperature and transistor characteristics. What improvement in gain stability results when 10 per cent of the out-of-phase output signal is fed back (i.e., the feedback factor $B = -0.10$)?

Solution:

$$G'_v = \frac{G_v}{1 - BG_v}$$

For $G_v = 50$

$$G'_v = \frac{50}{1 - (-0.10 \times 50)} = \frac{50}{1 + 5} = 8.33$$

Similarly, for $G_v = 200$

$$G'_v = \frac{200}{1 + 20} = 9.55$$

Hence for an open-loop gain variation of 300 per cent the closed-loop gain has a variation of

$$\frac{9.55 - 8.33}{8.33} \times 100\% = 14.6\%$$

The results of Example 5-1 are slightly misleading, since it would take at least two of the feedback stages to provide as much gain as the minimum gain of 50 for one nonfeedback stage.

EXAMPLE 5-2

What is the percentage of gain variation for two identical feedback stages connected back to back, as shown in Fig. 5-2?

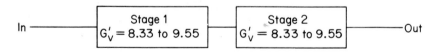

FIG. 5-2. Example 5-2

Solution:

When two amplifier stages are cascaded together, the total voltage gain equals the product of the two individual voltage gains. Hence the total G_v here will vary from a minimum of $(8.33)^2$ up to a maximum of $(9.55)^2$, or from about 70 up to 90. Therefore,

$$\% \text{ variation} = \frac{90 - 70}{70} \times 100\%$$

$$\simeq 28\%$$

Therefore, in practice, the use of a feedback amplifier of at least equivalent voltage gain as the amplifier not incorporating feedback, in this case, yields a variation reduction of from 300 down to 28 per cent. Note that the gain variation for two identical stages is approximately double the gain variation for one stage as calculated in Example 5-1.

5-3 Emitter Feedback Resistor

Figure 5-3 illustrates a general form for the CE amplifier. It incorporates no feedback and is plagued by stability and gain variation problems.

EXAMPLE 5-3

Consider the bias stability problem for the amplifier of Fig. 5-3 when designed for a nominal transistor h_{fe} of 100, but the h_{fe} can actually vary between 30 and 180 total for temperature and device-to-device variations.

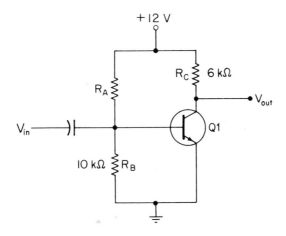

FIG. 5-3. Common-emitter amplifier—no feedback

Solution:
The first step is to solve for a value of R_A to provide a 6 V bias at $Q1$'s collector. Recall that operation at one-half the supply voltage allows for the maximum output signal, as well as generally permitting operation in the most linear portion of a transistor's output characteristics. This then minimizes distortion:

$$I_C = \frac{12 \text{ V} - 6 \text{ V}}{6 \text{ k}\Omega} = 1 \text{ mA}$$

$$h_{ib} \simeq \frac{0.026}{0.001} = 26 \, \Omega$$

Therefore, when $h_{fe} = 100$

$$R_{in} = 101 \times 26\ \Omega = 2.6\ \text{k}\Omega$$

If we assume a 0.7 V base–emitter forward bias, then by voltage divider action,

$$V_B = \frac{V_{CC}(R_B \| R_{in})}{R_A + (R_B \| R_{in})}$$

or

$$V_B = 12\ \text{V} \times \frac{10\ \text{k}\Omega \| 2.6\ \text{k}\Omega}{R_A + 10\ \text{k}\Omega \| 2.6\ \text{k}\Omega}$$

$$0.7\ \text{V} = \frac{12\ \text{V} \times 2.06\ \text{k}\Omega}{R_A + 2.06\ \text{k}\Omega}$$

$$R_A = 33.4\ \text{k}\Omega$$

We are now in a position to solve for the Q-point variation due to min–max h_{fe} values.

The R_{in} will be about the same in all three cases because of two opposite effects. Note that

$$R_{in} = \frac{h'_{fe} \times 0.026}{I_E} = h'_{fe} \times h_{ib}$$

As h_{fe} goes down, I_E tends to go down, and a counterbalancing effect takes place. This is a fairly crude approximation but adequate for this analysis. If R_{in} is constant, then I_B will be nearly constant; hence when $h_{fe} = 30$, I_C equals roughly $\frac{30}{100}$ of I_C when h_{fe} was 100. Hence when $h_{fe} = 30$

$$V_C = V_{CC} - (1\ \text{mA} \times \tfrac{30}{100} \times 6\ \text{k}\Omega)$$
$$= 12\ \text{V} - 1.8\ \text{V}$$
$$= 10.2\ \text{V}$$

The transistor is nearly cut off and the amplifier is useless unless the ac signal being amplified is extremely small.

Now when h_{fe} is 180, by similar logic,

$$V_C = V_{CC} - I_C \times \tfrac{180}{100} \times R_C$$
$$= 12\ \text{V} - 1\ \text{mA} \times 1.8 \times 6\ \text{k}\Omega$$
$$= 12\ \text{V} - 10.8\ \text{V} = 1.2\ \text{V}$$

Clearly the transistor is almost, if not fully, saturated with little hope of proper functioning.

Example 5-3 clearly shows the futility of using a CE amplifier without feedback because of Q-point shift. The variable voltage gain offered also prevents practical use of this circuit.

EXAMPLE 5-4

Solve for A_v when $h_{fe} = 30$, 100, and 200 in Example 5-3. At $h_{fe} = 100$,

$$A_v \simeq \frac{R_C}{h_{ib}} = \frac{6 \text{ k}\Omega}{26 \text{ }\Omega} = 230 \qquad (4\text{-}18)$$

At $h_{fe} = 30$,

$$h_{ib} = \frac{0.026 \text{ }\Omega}{10.2 \text{ V}/6 \text{ k}\Omega} = 15.2 \text{ }\Omega$$

$$A_v = \frac{6 \text{ k}\Omega}{15.2 \text{ }\Omega} \simeq 390$$

and at $h_{fe} = 200$,

$$h_{ib} = \frac{0.026 \text{ }\Omega}{1.2 \text{ V}/6 \text{ k}\Omega} = 130 \text{ }\Omega$$

$$A_v \simeq \frac{6 \text{ k}\Omega}{130 \text{ }\Omega} = 46$$

Hence the total voltage gain range is from 46 up to 390.

The circuit gain is clearly too highly variable and needs some negative feedback. The addition of a resistor in the emitter circuit will perform this function for us. To get a feel for how and why this feedback takes place, refer to Fig. 5-4. The biasing resistors are not shown in this circuit since they

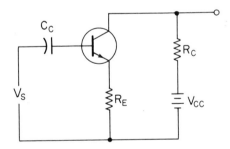

FIG. 5-4. Simplified CE feedback amplifier

have no effect on this discussion. In this circuit, the ac emitter current i_e is nearly equal to the ac output current i_c. The ac input voltage loop includes V_s, V_{be} (the input voltage determining $Q1$'s conduction), and V_e, where $V_e = i_e R_E$. The coupling capacitor is assumed to be a short circuit to the ac signal. As the instantaneous value of V_s increases, i_e increases, and hence V_e increases. Therefore, by Kirchhoff's voltage law, the transistor input voltage, V_{be}, does not increase as much as it would have without the emitter feedback resistor, R_E, in the circuit. Consequently, the voltage gain of this circuit has clearly been reduced by a process of negative feedback.

The stabilizing effect on the Q-point is evident in the following fashion: If h_{fe} were to increase due to a temperature change, the dc collector current would tend to increase, thus causing a Q-point drift. But if I_C increases, so does I_E; hence V_{R_E} increases. This increase in V_{R_E} causes the dc forward bias (V_{BE}) on the transistor to drop, since more dc drop across R_E leaves less voltage available for V_{BE}. Hence the conduction is reduced, and a strong stabilizing effect on I_C has taken place.

The amount of ac fed-back voltage is $i_e R_E$, and the ac output voltage is $i_c R_C$. The feedback factor B is therefore

$$B = -\frac{i_e R_E}{i_c R_C}$$
$$\simeq -\frac{R_E}{R_C} \tag{5-6}$$

The performance of this feedback amplifier is predicted by the general feedback equation developed earlier, Eq. (5-4).

5-4 Gain–Impedance Relations

A simplified equation for the gain of the CE amplifier with the feedback resistor is easily accomplished. Refer to Fig. 5-5 as an aid to understanding

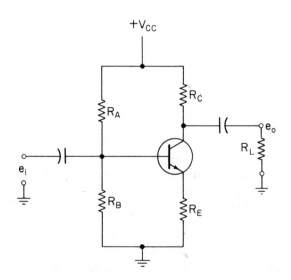

FIG. 5-5. Common-emitter amplifier with emitter feedback

the development of this equation. The voltage gain will equal e_o/e_i for this amplifier. The ac voltage across the load e_o is

$$e_o = i_c(R_L \| R_C) \tag{5-7}$$

If we assume that the ac emitter current is equal to the ac collector current and that the ac voltage across R_E follows the input voltage, we may write

$$e_i = i_c R_E \tag{5-8}$$

The voltage gain of this stage follows:

$$G_v \simeq \frac{e_o}{e_i} \simeq \frac{i_c(R_L \| R_C)}{i_c R_E} = \frac{R_L \| R_C}{R_E} \tag{5-9}$$

This is an extremely useful and practical result. When the proper relations between resistor values in Fig. 5-5 are maintained, this circuit yields a voltage gain determined almost exclusively by resistor values instead of transistor characteristics. If no external ac load resistor is used, the approximate voltage gain is then simply R_C/R_E.

The input impedance, Z_{in}, of this amplifier is also easy to approximate. From Chapter 4 we know that R_{in}, the input impedance looking into the transistor's base, equals $r_b + h'_{fe}r_e$ when no feedback resistor is used. It follows that

$$R_{in} = r_b + h'_{fe}(r_e + R_E) \tag{5-10}$$

when an external emitter resistor is added. This can be algebraically changed to

$$R_{in} = h'_{fe}(h_{ib} + R_E) \tag{5-11}$$

In most amplifiers R_E is from 200 to 2000 Ω, and h_{ib} is at least 10 times smaller (an order of magnitude less). Hence we are left with

$$R_{in} \simeq h'_{fe}R_E \tag{5-12}$$

If we now include the effects of the biasing resistors on the overall input impedance, Z_{in}, of this circuit, we see that

$$Z_{in} = R_A \| R_B \| h'_{fe}R_E \tag{5-13}$$

Typical values for these parameters are $R_A = 100$ kΩ, $R_B = 5$ kΩ, and $h'_{fe}R_E = 50$ kΩ; therefore, we can approximate the circuit input impedance Z_{in} as

$$Z_{in} \simeq R_B \tag{5-14}$$

Thus another important parameter of this circuit can usually be approximated by a resistor value. Of special significance is the fact that the input impedance of the feedback amplifier is usually an order of magnitude higher than the nonfeedback version. This is another desirable result of this feedback, since a higher input impedance means less loading effects on the voltage source. In Section 5-8 feedback that reduces Z_{in} will be treated.

The current gain of this circuit is easily derived by circuit analysis or the TGIR. Since $G_i = G_v(R_I/R_L)$, where in this case $R_L = R_L \| R_C$, then

$$G_i = \frac{R_L \| R_C}{R_E} \times \frac{R_B}{R_L \| R_C} = \frac{R_B}{R_E} \qquad (5\text{-}15)$$

Once again an important characteristic of this amplifier is determined by resistor values. This result is often referred to as the *stability factor* or *S-factor* of an amplifier:

$$S = G_i = \frac{R_B}{R_E} \qquad (5\text{-}16)$$

The lower the S-factor the greater will be the stability of the amplifier. S-factors of from 5 to 20 are typical, but S should always be an order of magnitude smaller than the minimum h_{fe} transistor to be used in the design. A ratio of at least 5:1 is desirable. If the S-factor is not always at least five times smaller than h_{fe}, the approximations we have made for G_v, G_i, and R_I will start to have appreciable error.

The feedback CE circuit (Fig. 5-5) yields voltage and current gains of from 5 to 20 as compared to a current gain of h_{fe} and a voltage gain of several hundred in the nonfeedback circuit. The sacrifice in voltage gain is almost always desirable due to the resultant stability and predictability. Figure 5-6

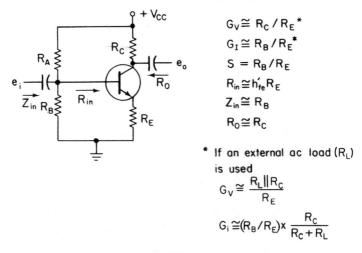

FIG. 5-6. Emitter feedback amplifier summary

summarizes the important results of this section. Note that the output impedance approximation is given as R_C. In reality it is equal to R_C in parallel with the impedance seen looking into the reverse-biased collector. The collector impedance is thus very high, justifying the given approximation. If an external ac load is connected (as shown dashed in Fig. 5-6), it appears in parallel with R_C, and G_v is then approximated as $(R_C \| R_L)/R_E$.

5-5 Emitter Feedback Biasing

Unfortunately, all amplifying devices require dc biasing so that proper operation can take place. The calculations we have previously made for biasing nonfeedback CE amplifiers are, from a practical application standpoint, useless. A CE amplifier must use feedback. The validity of this conclusion is illustrated by the variations in Q-point in Example 5-3. We shall now repeat that example with an emitter feedback resistor of 1.5 kΩ added.

EXAMPLE 5-5

For the circuit of Fig. 5-7 determine the proper value for R_A to provide a 6 V bias at the collector of $Q1$ when $h'_{fe} = 100$. Then determine the change in this Q-point for an h'_{fe} spread of 30–180.

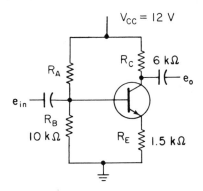

FIG. 5-7. Example 5-6

Solution:
If $V_C = 6$ V, then $I_C = 1$ mA and $I_E \simeq 1$ mA:

$$V_E = (1 \text{ mA})(1.5 \text{ k}\Omega) = 1.5 \text{ V}$$
$$V_B = 1.5 \text{ V} + 0.7 \text{ V} = 2.2 \text{ V}$$

98 Single-Stage Feedback Amplifiers

Now $R_{in} = h'_{fe}R_E \simeq 150 \text{ k}\Omega$, so by voltage divider action

$$2.2 \text{ V} = 12 \text{ V} \times \frac{(10 \text{ k}\Omega \| 150 \text{ k}\Omega)}{R_A + (10 \text{ k}\Omega \| 150 \text{ k}\Omega)}$$

$$= 12 \text{ V} \times \frac{9.4 \text{ k}\Omega}{R_A + 9.4 \text{ k}\Omega}$$

$$R_A = 41.5 \text{ k}\Omega$$

If h'_{fe} dropped to 30,

$$R_{in} = 30 \times 1.5 \text{ k}\Omega = 45 \text{ k}\Omega$$

$$V_B = 12 \text{ V} \times \frac{(10 \text{ k}\Omega \| 45 \text{ k}\Omega)}{41.5 \text{ k}\Omega + (10 \text{ k}\Omega \| 45 \text{ k}\Omega)}$$

$$\simeq 2 \text{ V}$$

$$V_E = 2 \text{ V} - 0.7 \text{ V} = 1.3 \text{ V}$$

$$I_E = \frac{1.3 \text{ V}}{1.5 \text{ k}\Omega} = 0.87 \text{ mA} \simeq I_C$$

$$V_C = 12 \text{ V} - (0.87 \text{ mA})(6 \text{ k}\Omega)$$

$$= 6.8 \text{ V}$$

So the new Q-point shifted from 6 to 6.8 V, whereas the nonfeedback circuit of Example 5-3 shifted from 6 to 10.2 V under the same conditions. The Q-point shift when h'_{fe} is at the maximum limit of 180 is calculated the same way as for the minimum-limit calculation:

$$R_{in} = h'_{fe}R_E \simeq 270 \text{ k}\Omega$$

$$V_B = 12 \text{ V} \times \frac{(270 \text{ k}\Omega \| 10 \text{ k}\Omega)}{41.5 \text{ k}\Omega + 9.6 \text{ k}\Omega} = 12 \text{ V} \times \frac{9.6 \text{ k}\Omega}{41.5 \text{ k}\Omega + 9.6 \text{ k}\Omega} = 2.25 \text{ V}$$

$$V_E = 2.25 \text{ V} - 0.7 \text{ V} = 1.55 \text{ V}$$

$$I_E = \frac{1.55 \text{ V}}{1.5 \text{ k}\Omega} = 1.03 \text{ mA} \simeq I_C$$

$$V_C = 12 \text{ V} - (1.03 \text{ mA})(6 \text{ k}\Omega)$$

$$= 5.8 \text{ V}$$

Thus the Q-point shift of this stabilized circuit is only from 6.8 to 5.8 V for an h'_{fe} spread of 30 to 180. The circuit is now practical and useful.

In practice, the value of R_A in Example 5-5 can successfully be determined by ignoring the shunting effect R_{in} has on R_B. In that case

$$2.2 \text{ V} = 12 \text{ V} \times \frac{R_B}{R_B + R_A} = 12 \text{ V} \times \frac{10 \text{ k}\Omega}{10 \text{ k}\Omega + R_A}$$

$$R_A = 44.5 \text{ k}\Omega$$

which is quite close to the result from the exact solution of 41.5 kΩ. When

$h'_{fe} = 100$, the error in bias point as a result of this simplified solution can be determined as follows:

$$V_B = 12 \text{ V} \times \frac{(10 \text{ k}\Omega \parallel 50 \text{ k}\Omega)}{44.5 \text{ k}\Omega + (10 \text{ k}\Omega \parallel 50 \text{ k}\Omega)} \simeq 2.1 \text{ V}$$

$$V_E = 2.1 \text{ V} - 0.7 \text{ V} = 1.4 \text{ V}$$

$$I_E = \frac{1.4 \text{ V}}{1.5 \text{ k}\Omega} = 0.935 \text{ mA} \simeq I_C$$

$$V_C = 12 \text{ V} - (0.935 \text{ mA} \times 6 \text{ k}\Omega)$$
$$= 6.4 \text{ V}$$

This result is close enough to the goal of 6 V to justify the use of this approximation.

If the amplifier is to be a one-of-a-kind variety, the best way to determine R_A is to breadboard the circuit in the laboratory and experimentally determine the value that gives the desired Q-point. The reason for this is that only by experimental procedures can the total complex problem of biasing be reckoned with, unless a circuit study well beyond the scope of this textbook is made.

In practice, the values of R_B, R_C, and R_E can be scaled up or down so long as the current gain R_B/R_E and voltage gain are kept below 20. Exceeding this approximate limit means that a low amount of feedback is in effect, and insufficient stability will result. Keep in mind that this gain limit of 20 is too high if low h_{fe} transistors are being used, and could be higher if only very high h_{fe} units are being used. A good rule of thumb, as stated previously, is to use an S-factor no greater than one fifth of the transistor's h_{fe}.

Recall from Eq. (5-6) that the feedback factor B for this circuit is $-R_E/R_C$. Therefore, our example problem has

$$B = -1.5 \text{ k}\Omega/6 \text{ k}\Omega = -0.25$$

Applying the general feedback equation, Eq. (5-4), we have

$$G'_v = \frac{G_v}{1 - BG_v}$$

where G'_v is the closed-loop gain (with feedback) and G_v is the open-loop gain. Therefore,

$$G_v = \frac{R_C}{h_{ib}} = \frac{6 \text{ k}\Omega}{0.026/1 \text{ mA}}$$
$$= 230$$

and

$$G'_v = \frac{230}{1 - (-0.25 \times 230)} = \frac{230}{1 + 57.5}$$
$$= 4$$

This is the same result predicted by our approximate gain formula R_C/R_E, and hence agreement between these two equations exists as it should.

5-6 Emitter Bypass Capacitor

In some applications the emitter feedback resistor is shunted by a large-valued (therefore usually electrolytic) capacitor. To the Q-point (a dc parameter) the capacitor appears as an open circuit, and hence has no effect. Thus Q-point stability is unaffected and will remain satisfactory. However, to the ac signal we'll consider the capacitor to be a short, and hence the ac voltage gain should go up since the ac feedback path has been eliminated. In fact, G_v is now once again the same as predicted by the nonfeedback CE amplifier:

$$G_v = \frac{R_C}{h_{ib}} \qquad (5\text{-}17)$$

(See Fig. 5-8.) Thus this technique offers a tremendous increase in gain, while still providing the necessary Q-point stability. This all appears well and good,

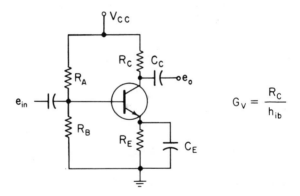

FIG. 5-8. Emitter bypass

but two conditions must be considered:

1. The current gain of this amplifier is now roughly equal to h_{fe}, and hence is dependent on transistor characteristics and therefore unpredictable. The input impedance (Z_{in}) is now equal to $R_A \| R_B \| h'_{fe} h_{ib}$ with $h'_{fe} h_{ib}$ normally the smallest of the three, and hence the controlling factor. This means Z_{in} is dependent on h_{fe} also.
2. As the signal frequency goes down, the capacitor (C_E) stops appearing as a short as its reactance starts increasing. The voltage gain then equals $R_C/(h_{ib} \| -jX_c)$; hence the voltage gain starts falling off.

Good low-frequency response requires large values of capacitance (200 μF is not uncommon). (See Section 8-3 for details on calculating the cutoff frequency.) This requirement eliminates the use of this configuration from most new amplifier designs for economic reasons. High-valued capacitors are expensive and bulky compared with making up for the decreased gain of nonbypassed circuits by adding on an extra stage. In previous years transistors were more expensive than electrolytic capacitors, but now the reverse is true. This is especially true of integrated circuits in which capacitors of greater than about 0.01 μF are virtually impossible to attain.

EXAMPLE 5-6

For the circuit of Fig. 5-9, determine the proper value for R_A, the voltage gain, and the current gain.

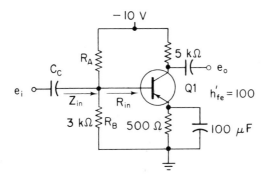

FIG. 5-9. Example 5-7

Solution:
First find $I_C = 1$ mA for a -5 V bias; then $V_E = 1$ mA $\times 0.5$ kΩ $= -0.5$ V, and $V_B = -0.5$ V $- 0.7$ V $= -1.2$ V:

$$-1.2 \text{ V} \simeq -10 \text{ V} \frac{3 \text{ k}\Omega}{R_A + 3 \text{ k}\Omega}$$

$$R_A = 22 \text{ k}\Omega$$

$$G_v = \frac{R_L}{h_{ib}} \quad \text{and} \quad h_{ib} = \frac{0.026}{1 \text{ mA}} = 26 \text{ } \Omega$$

$$= \frac{5 \text{ k}\Omega}{26 \text{ } \Omega} = 192$$

Then $G_i = G_v(R_I/R_L)$ and

$Z_{in} = 22 \text{ k}\Omega \| 3 \text{ k}\Omega \| R_{in} = 22 \text{ k}\Omega \| 3 \text{ k}\Omega \| (h'_{fe} \times 26 \text{ } \Omega) = 22 \text{ k}\Omega \| 3 \text{ k}\Omega \| 2.6 \text{ k}\Omega$

$Z_{in} = 1.3 \text{ k}\Omega$

$$G_i = 192 \times \frac{1.3 \text{ k}\Omega}{5 \text{ k}\Omega} = 50$$

Notice that the current gain of this amplifier is only half $Q1$'s h_{fe}. The reason for this is that only half the ac input current flowing through C_c reaches the base lead of $Q1$. From Kirchhoff's current law,

$$i_b = i_I \frac{R_A \| R_B}{(R_A \| R_B) + R_{in}} = i_I \frac{22 \text{ k}\Omega \| 3 \text{ k}\Omega}{22 \text{ k}\Omega \| 3 \text{ k}\Omega + 2.6 \text{ k}\Omega}$$

$$= \frac{i_I \times 2.6 \text{ k}\Omega}{2.6 \text{ k}\Omega + 2.6 \text{ k}\Omega} = \frac{i_I}{2}$$

In many instances a technique known as partial bypassing is employed as a compromise. It offers better low-frequency response (for the same size bypass capacitor), but the tradeoff is less voltage gain. Figure 5-10 shows

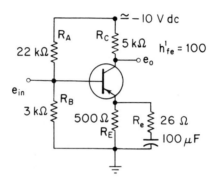

FIG. 5-10. Partial emitter bypass

the circuit of Example 5-6 with a partial bypass. A 26 Ω resistor has been added in series with the 100 μF bypass capacitor.

EXAMPLE 5-7

Repeat Example 5-6 for the partial bypass condition of Fig. 5-10.

Solution:

The dc Q-point conditions are not changed from Example 5-6 since the only circuit change (the 26 Ω partial bypass resistor) has been added in series with an open circuit (the 100 μF capacitor) to the dc signal. Therefore, h_{ib} is still 26 Ω as before and the 22 kΩ value of R_A is still valid; but the ac voltage gain calculation must take into account the added emitter resistance:

$$G_v = \frac{R_L}{h_{ib} + R_e}$$

where R_e in this case is 26 Ω ∥ 500 Ω ≃ 26 Ω and R_L is the 5-kΩ collector resistor:

$$G_v \simeq \frac{5 \text{ k}\Omega}{26 \text{ }\Omega + 26 \text{ }\Omega} = 96$$

$$G_i = G_v \frac{R_I}{R_L}$$

and

$$Z_{in} = 22 \text{ k}\Omega \,\|\, 3 \text{ k}\Omega \,\|\, [h'_{fe}(h_{ib} + R_e)]$$
$$= 22 \text{ k}\Omega \,\|\, 3 \text{ k}\Omega \,\|\, 5.2 \text{ k}\Omega$$
$$= 1.75 \text{ k}\Omega$$

$$G_i = 96 \frac{1.75 \text{ k}\Omega}{5 \text{ k}\Omega} = 33.5$$

The voltage gain of this amplifier is one half the full bypass circuit, but it will respond down to lower frequencies.

5-7 Collector Feedback—CE Amplifier

Another form of feedback that is commonly used is known as *collector feedback*. It is also referred to as shunt feedback, because a portion of the output is fed back in *shunt* with the input. Figure 5-11 should make the reason

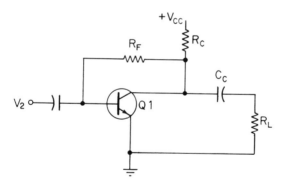

FIG. 5-11. Common-emitter amplifier—collector feedback

for the name collector feedback apparent—a resistor, R_F, is connected from the collector, the output, back to the input. The feedback resistor also functions in the role of supplying direct current to $Q1$'s base in order to set up the bias point. Although this circuit form is not used nearly as much as the one previously studied, it does merit further study here.

To understand the stabilizing effect that R_F has on the Q-point, assume that h_{fe} has drifted higher. This causes an increase in collector current, which causes the dc collector voltage to drop. With less dc collector voltage applied to R_F, the base current will decrease and hence compensate for the increase in h_{fe}.

We need to derive an expression to solve for the circuit values that will allow for a proper bias point. To do that, we write serveral loop equations, noting that the transistor's base and collector current flow through R_C. Therefore,

$$V_C = V_{CC} - (I_C + I_B)R_C \simeq V_{CC} - I_C R_C$$

and also

$$V_C = I_B R_F + V_{BE}$$

Equating these two expressions,

$$I_B R_F + V_{BE} = V_{CC} - I_C R_C$$

and, noting that $I_B = I_C/h_{fe}$,

$$\frac{I_C R_F}{h_{fe}} + V_{BE} = V_{CC} - I_C R_C$$

We then solve for I_C:

$$I_C = \frac{V_{CC} - V_{BE}}{R_C + (R_F/h_{fe})}$$

$$\simeq \frac{V_{CC}}{R_C + R_F/h_{fe}} \qquad (5\text{-}18)$$

since V_{BE} can usually be neglected with respect to V_{CC}. Equation (5-18) gives us the necessary information to analyze the dc aspects of the collector feedback circuit.

EXAMPLE 5-8

For the circuit shown in Fig. 5-12, determine the value of R_F to provide proper biasing.

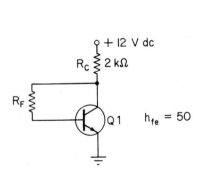

FIG. 5-12. Example 5-8

Solution:
Simply plug the known values into Eq. (5-18) and solve for the necessary value of R_F. With a bias of 6 V, R_C should have 6 V across it, and hence a current flow of

3 mA, which is almost equal to I_C:

$$3 \text{ mA} = \frac{12 \text{ V}}{2 \text{ k}\Omega + (R_F/50)}$$

$$R_F = 100 \text{ k}\Omega$$

Alternative Solution:

Instead of using the developed formula, perhaps more insight into this circuit could be gained by simple analysis. As before, we know that the proper bias point for the collector is 6 V dc. Therefore, if we neglect the base–emitter voltage we can see that R_F will have 6 V across it. The current through R_F is I_B, and $I_B = I_C/h_{fe}$:

$$I_B = \frac{6 \text{ V}/2 \text{ k}\Omega}{h_{fe}} = \frac{3 \text{ mA}}{50} = 0.06 \text{ mA}$$

$$R_F = \frac{6 \text{ V}}{0.06 \text{ mA}} = 100 \text{ k}\Omega$$

This method of solution eliminates the need for reference to the provided formula and forces the student to analyze the circuit.

5-8 Collector Feedback Input Impedance and Gain

We shall now determine the input impedance and gain for this type of circuit. Reference to Fig. 5-13 will help in solving for R'_{in} (the impedance

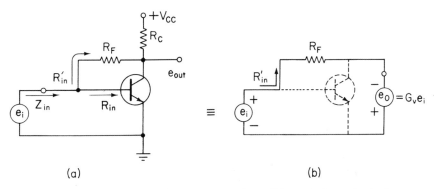

FIG. 5-13. (a) Collector feedback amplifier; (b) equivalent circuit for finding R_I

looking into R_F). Once we have R'_{in} we know that $R_{in} = h'_{fe}h_{ib}$, and we can solve for the overall input impedance, Z_{in}, as being the parallel combination of R'_{in} and R_{in}. Taking the sum of voltages in the outer loop of Fig. 5-13b, we find that R_F has $(e_I + G'_v e_I)$ volts across it, recalling that phase inversion exists from the input to output of a CE amplifier. Keep in mind that G'_v is the actual voltage gain of the transistor with the effects of feedback included:

$$V_{R_F} = e_I + G'_V e_I = (1 + G'_V)e_I$$

where G'_V is the voltage gain of this amplifier including the effects of negative feedback. Therefore,

$$\begin{aligned} R'_{in} &= \frac{e_I}{i_I} = \frac{e_I}{V_{R_F}/R_F} \\ &= \frac{e_I}{(1 + G'_V)e_I/R_F} \quad (5\text{-}19) \\ &= \frac{R_F}{1 + G'_V} \end{aligned}$$

Equation (5-19) shows that looking into a resistance connected from the output to the input of a phase-inverting amplifier results in an effective reduction of impedance by a factor of 1 plus the voltage gain of that amplifier. This is an interesting effect, which will also be important in our study of the high-frequency response of amplifiers. This effect also causes multiplication of the inherent base–collector junction capacitance by this $(1 + G'_V)$ factor. This then is often the limiting factor on the high-frequency response of an amplifier. We will look at this effect in detail in Chapter 8.

To determine the voltage gain of this amplifier, the general feedback equation should be used. The gain of this amplifier without feedback, G_V, is simply equal to R_C/h_{ib}. The feedback factor, B, is

$$B = -\frac{R_C}{R_F} \quad (5\text{-}20)$$

Thus the actual voltage gain is

$$\begin{aligned} G'_V &= \frac{G_V}{1 - BG_V} \quad (5\text{-}4) \\ &= \frac{R_E/h_{ib}}{1 - [-(R_C/R_F) \times (R_C/h_{ib})]} \\ &= \frac{R_C}{h_{ib} + (R_C^2/R_F)} \quad (5\text{-}21) \end{aligned}$$

Once the actual voltage gain is known, the current gain can best be calculated using the TGIR.

EXAMPLE 5-9

Find the input impedance, voltage gain, and current gain for the amplifier used in Example 5-8.

Solution:
The input impedance equals $h'_{fe}h_{ib}$ in parallel with $R_F/(1 + G'_V)$:

$$h_{ib} = \frac{0.026}{0.003 \text{ A}} = 8.67 \text{ }\Omega$$

and the open-loop voltage gain equals R_C/h_{ib}.

$$G_V = \frac{2 \text{ k}\Omega}{8.67 \text{ }\Omega} = 230$$

Therefore

$$G'_V = \frac{R_C}{h_{ib} + (R_C^2/R_F)}$$

$$= \frac{2 \text{ k}\Omega}{8.67 \text{ }\Omega + [(2 \text{ k}\Omega)^2/100 \text{ k}\Omega]} = \frac{2 \text{ k}\Omega}{8.67 \text{ }\Omega + 40 \text{ }\Omega} \quad (5\text{-}21)$$

$$= 41$$

Now the input impedance Z_{in} can be calculated as

$$Z_{in} = (51 \times 8.67 \text{ }\Omega) \| \frac{100 \text{ k}\Omega}{42}$$

$$= 441 \text{ }\Omega \| 2.38 \text{ k}\Omega$$

$$= 375 \text{ }\Omega$$

Now calculate the current gain using the TGIR as

$$G_i = 41 \frac{375 \text{ }\Omega}{2 \text{ k}\Omega}$$

$$= 7.7$$

Example 5-9 illustrates the fact that the input impedance for this circuit has been decreased by the addition of feedback. This is a characteristic of shunt feedback—the input impedance is decreased, whereas the effect with the emitter feedback resistor (series feedback) is to increase the input impedance. Since the bipolar transistor is basically a low-impedance device to begin with, this lowering of input impedance by shunt feedback is a disadvantage of the collector-feedback method. This explains its more limited use with respect to emitter feedback, because it can only be driven by low-impedance sources. The lower output impedance that exists with collector feedback is, however, an advantage that somewhat counterbalances the low input impedance effect.

5-9 Voltage Divider Feedback

One last circuit will complete our study of single-stage bipolar transistor feedback amplifiers. Other feedback schemes exist, but the ones developed in this chapter are those most frequently encountered, and an understanding of them will make analysis of virtually any other scheme possible.

Consider the system shown in Fig. 5-14. The voltage feedback ratio B

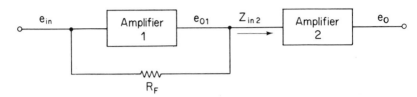

FIG. 5-14. Voltage divider feedback

for amplifier 1 is equal, by voltage divider action, to

$$B = -\frac{R_F}{R_F + Z_{in_2}}$$

since the fraction $R_F/(R_F + Z_{in_2})$ is the percentage of e_o fed back to the first amplifier's input. Amplifier 1 must be a phase-inverting amplifier so that the fed-back signal is out of phase with the input signal to introduce the necessary negative feedback.

EXAMPLE 5-10

If the open-loop voltage gain of each stage in Fig. 5-14 is 100, solve for the overall voltage gain of this system given $Z_{in_2} = 50 \text{ k}\Omega$ and $R_F = 2 \text{ k}\Omega$.

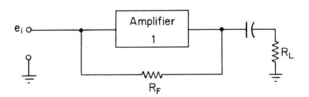

FIG. 5-15. Example 5-10

Solution:

$$G'_{V1} = \frac{G_{V1}}{1 - BG_{V1}} = \frac{100}{1 - \left(-\frac{2 \text{ k}\Omega}{50 \text{ k}\Omega}\right)(100)}$$

$$= 20$$

Therefore, the overall system voltage gain (G_{voa}) is

$$G_{voa} = G'_{V1} \times G_{V2}$$
$$= 20 \times 100$$
$$= 2000$$

PROBLEMS

1. Mathematically derive Eq. (5-4) from Eq. (5-3).
2. Why is the result of Eq. (5-5) such a desirable one?
3. An amplifier design yields a total voltage gain variation of from 25 to 300 without feedback. Calculate the improvement in percentage of variation when 10 per cent feedback is incorporated. How many of these feedback stages would be needed to provide at least as much gain as the original amplifier?
4. (a) Repeat Problem 3 for 5 per cent feedback.
   (b) Repeat Problem 3 for 1 per cent feedback.
5. The value of R_A in Fig. P5-5 is to be determined by designing for a bias

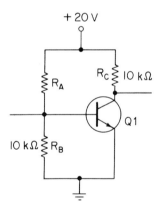

FIG. P5-5.

voltage of 10 V at the collector of $Q1$ for a nominal h'_{fe} of 80. The total possible h'_{fe} variation for temperature and device-to-device variations is from 25 to 150. Determine the possible Q-point and voltage gain variations.

6. Derive Eq. (5-11) from Eq. (5-10).
7. Develop Eq. (5-15) by circuit analysis instead of the TGIR.
8. Determine the input impedance and voltage gain for the circuit shown in Fig. P5-5. Assume h'_{Fe} 100 and $V_C = 10\ V$.
9. Calculate the current gain for the circuit shown in Fig. P5-5.
10. (a) Determine the collector Q-point for the amplifier shown in Fig. P5-10.

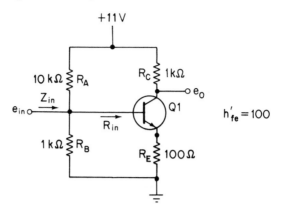

FIG. P5-10.

(b) Determine G_v, G_i, S, Z_{in}, and the output impedance for this amplifier.

11. To the circuit in Fig. P5-10 add a 50 μF capacitor across the emitter resistor. Determine G_v, G_i, and Z_{in}.

12. Repeat Problem 11 with a 15 Ω partial bypass resistor in series with the bypass capacitor.

13. For the circuit of Fig. P5-13 solve for R_F to provide proper biasing ($V_C = 8$ V).

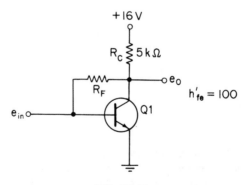

FIG. P5-13.

14. For the circuit of Fig. P5-13 determine the input impedance, voltage gain, and current gain.

15. For Fig. P5-15 determine the ac voltage across the load R_L.

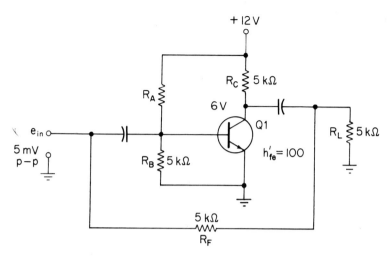

FIG. P5-15.

6

Field-Effect-Transistor Amplifiers

6-1 Introduction

The field-effect transistor (FET) has come into its own in today's electronic circuits. Although basic research on this device was taking place at or before that of the bipolar transistor, it did not come into general usage until the 1960s. Mass production difficulties held up its commercial availability, and even now some of these same problems, to a lesser extent, are responsible for a wide variance in its gain parameters.

Field-effect transistors come in two general categories, the junction field-effect transistor (JFET) and the metal oxide semiconductor field-effect transistor (MOSFET). The MOSFET is sometimes referred to as the IGFET with the IG standing for insulated gate. The meaning of all this will become apparent as our discussion progresses. Field-effect transistors are also sometimes referred to as *unipolar* transistors since conduction is by majority charge carriers only. This is in contrast to *bipolar* junction transistors in which both majority and minority charge carriers are involved in the conduction process.

The major advantage of FETs over bipolar transistors is their extremely high input impedance. Input impedances of 10^8 to 10^{14} Ω are available as compared to 10^2 to 10^6 Ω for bipolar transistors—a considerable advantage for many applications. The FET also offers advantages in noise characteristics when working from high impedances, temperature effects, and simpler biasing techniques. The biasing of MOSFETs operating in the enhancement mode is uniquely simple. One final advantage of extreme importance to today's circuits is the fabrication simplicity and lower cost of integrated circuits using MOSFETs. At this point, the student is entitled to ask why haven't FETs completely supplanted bipolar transistors? The bipolar transistor is capable of higher voltage gains, better high-frequency response, and greater power dissipations. The selection between the use of unipolar or bipolar transistors is often a complex technical–economic decision depending on many factors, with many of today's circuits being a combination of both varieties.

6-2 JFET Theory of Operation

The JFET is formed as an *n*-type or *p*-type *channel* and with an opposite polarity semiconductor material known as the gate. The physical construction is shown in Fig. 6-1a. Figure 6-1b shows the appropriate symbols and

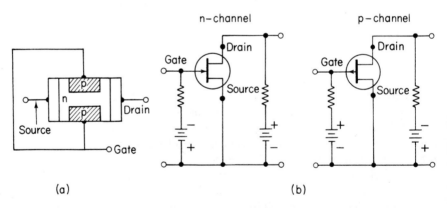

FIG. 6-1. (a) Junction-field-effect transistor structure *n*-channel; (b) JFET symbols and necessary voltage polarities

required voltage polarities for *n*-channel and *p*-channel operation. The names of the three leads of a FET are also noted. The drain and source leads are often interchangeable and are determined by the polarity of the supply voltage to which they are connected. The drain is the lead connected to the positive side of the battery for the *n*-channel device and to the negative side for the *p*-channel device.

A rough analogy exists between the JFET and bipolar transistor with respect to gate and base, drain and collector, and emitter and source. However, the JFET's conduction is controlled by the reverse bias voltage on the gate–source junction and the electric field it creates. The bipolar transistor's conduction control is by the base current in the forward-biased base–emitter junction. This accounts for the JFET's high input impedance, since its input (the gate) is a reverse-biased junction.

Figure 6-2 provides a curve of drain current, I_D, versus the drain-to-source voltage, V_{DS}. The bias voltage or voltage from gate to source, V_{GS}, is

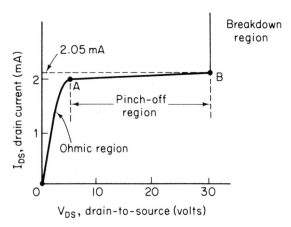

FIG. 6-2. Curve of I_D versus V_{DS} with $V_{GS} = 0$ for JFET

equal to zero in this instance. In the first area, from 0 to A, the *ohmic region*, the JFET acts as an ordinary resistor. The current increases linearly as the applied voltage is increased. This happens up to 1–3 V typically until the *pinch-off* region begins. At this point (from A to B) the reverse bias on the gate-to-channel *pn* junction has caused a *depletion region* in the channel next to the junction, as shown in Fig. 6-3. Until the reverse bias is large enough to

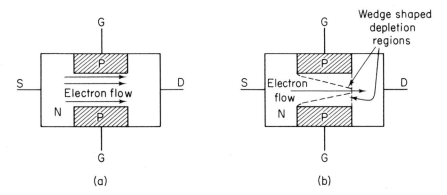

FIG. 6-3. Cross section of a JFET: (a) no bias; (b) biased

reach the pinch-off region, the depletion layers are not wide enough to meet. When enough bias is provided to allow the layers to touch, the channel is depleted of charge carriers and is *pinched off*. Be careful at this point, since "pinch off" *does not* mean "current off." Pinch off is an area of almost *constant* current flow for wide changes in V_{DS}. Refer to Fig. 6-2 and note that the curve is almost horizontal in this region from A to B. As the channel width is decreasing to zero, it tends to make the current flow decrease, which causes a drop in the voltage gradient along the channel. This then reduces the depletion width, which causes the current to increase. The final result is a stabilized value of current that is in a state of equilibrium.

The *dynamic drain resistance*, r_{ds}, is defined as the slope of the curve in the pinch-off region:

$$r_{ds} = \frac{\Delta V_{DS}}{\Delta I_D}\bigg|_{V_{GS}=\text{constant}} \qquad (6\text{-}1)$$

It is a very high value, and for Fig. 6-2 between points A and B is

$$r_{ds} \simeq \frac{30\text{ V} - 5\text{ V}}{2.05\text{ mA} - 2\text{ mA}} = \frac{25\text{ V}}{0.05\text{ mA}} = 500\text{ k}\Omega \qquad (6\text{-}1)$$

This is the ac resistance of the channel, since it is the resistance to changes in V_{DS}. On the other hand, the static, or dc, resistance of the channel at a given V_{DS} is a low resistance and is given simply by the ratio of V_{DS} to I_{DS}. Hence for Fig. 6-2, at $V_{DS} = 20$ V,

$$R_{DS} = \frac{V_{DS}}{I_D} \simeq \frac{20\text{ V}}{2\text{ mA}} = 10\text{ k}\Omega \qquad (6\text{-}2)$$

The breakdown region for voltages beyond point B is also shown in Fig. 6-2. Breakdown results when the reverse-biased gate–channel *pn* junction undergoes avalanche breakdown. It results in a rapid increase in current flow for a small change in V_{DS}. It is similar to the constant voltage region of a zener diode and is nondestructive if the FET's power rating is not exceeded. It is interesting to note that increasing values of V_{DS} cause the JFET to appear first as a resistor (ohmic region), then as a constant current source (pinch-off region), and finally as a constant voltage source (breakdown region).

In the preceding discussion, the gate and source were at the same potential ($V_{GS} = 0$). If a plot of I_D versus V_{DS} were made at different values of V_{GS}, the *drain characteristic curves*, as shown in Fig. 6-4, would result. It is customary to not show the breakdown (zener) region on these curves, since JFETs are seldom operated in this region. Notice that various (increasingly negative) bias potentials (V_{GS}) result in decreasing values of drain current in the pinch-off region. This is the basis for an FET's amplifying ability, and its *transcon-*

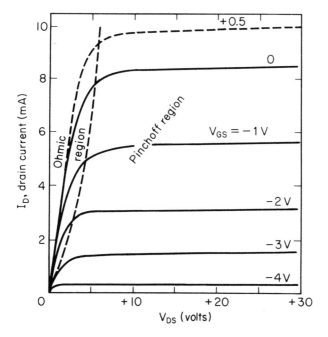

FIG. 6-4. Drain characteristic curves of a JFET

ductance (g_{fs}) is a measure of the change in drain current for a change in V_{GS}:

$$g_{fs} = \left.\frac{\Delta I_D}{\Delta V_{GS}}\right|_{V_{DS}=\text{constant}} \quad (6\text{-}3)$$

The transconductance is variable to a certain extent, as shown by the unequal spacing of I_D for 1 V changes in bias voltage in Fig. 6-4. The variation in g_{fs} from device to device for a specific type of FET is very high also, with max-min ratios of 3:1 typical. Since G_v for an FET amplifier is proportional to g_{fs}, negative feedback is necessary for most applications. The transconductance g_{fs} is termed the *transadmittance* y_{fs} by some manufacturers. Figure 6-1b shows a typical common-drain JFET amplifier, and the voltage gain of this FET without negative feedback, with the input at the gate and output at the drain, is

$$A_v = g_{fs}\frac{R_D r_{ds}}{R_D + r_{ds}} \quad (6\text{-}4)$$

which reduces to

$$A_v \simeq g_{fs} R_D \quad (6\text{-}5)$$

when $r_{ds} \gg R_D$, as is usually the case. If an external ac load resistor, R_L, is

118 *Field-Effect-Transistor Amplifiers*

used, the voltage gain is

$$G_v \cong g_{fs}(R_D \parallel R_L) \qquad (6\text{-}6)$$

The transconductance of the device represented by the characteristics of Fig. 6-4 for a change of V_{GS} from -1 to -2 V is

$$g_{fs} = \frac{\Delta I_D}{\Delta V_{GS}} \simeq \frac{5.5 \text{ mA} - 3 \text{ mA}}{1 \text{ V}} = 2500 \ \mu\mho \qquad (6\text{-}3)$$

Note also from Fig. 6-4 that operation with a small forward bias is permissible as long as the gate–channel junction is not allowed to conduct (become forward biased). Since FETs are normally made of silicon material, this would imply a positive voltage no greater than about 0.5 V. Conduction would of course mean a significant current flow and probably cause destruction of the device due to heating effects in the *pn* junction.

The JFET can either be *n*-channel or *p*-channel, but the *n*-channel is normally preferred. It offers conduction via electrons, whereas the *p*-channel device operates via hole conduction. Since electrons are more mobile than holes, the *n*-channel device allows a higher frequency of operation, all other factors being equal. It also turns out that *n*-channel devices introduce less noise into a circuit than do their *p*-channel counterparts. The *n*-channel characteristic curves of Fig. 6-4 are equally applicable to a *p*-channel device. However, all voltage polarities would have to be reversed.

Table 6-1 is a summary of the more important FET parameters and definitions. Unfortunately, manufacturers have not yet settled on universal standards for FET notations. The ones given in Table 6-1 are the more commonly accepted ones. As is usual, capitals are used for dc parameters and lowercase for ac. Single and double subscript notations are generally self-explanatory, but the triple subscripts require explanation. The third subscript *S* in Table 6-1 indicates that the terminal not designated by the first two subscripts is shorted to the terminal indicated by the second subscript. A third subscript *O* indicates that the unmentioned terminal is left open.

Shockley has shown that a fixed relationship exists between several important JFET parameters:

$$I_D \simeq I_{DSS}\left(1 - \frac{V_{GS}}{V_P}\right)^2 \qquad (6\text{-}7)$$

I_{DSS} is the drain current for zero gate voltage and V_P (the pinch-off voltage) is the value of V_{GS} required to effectively "turn off" the FET. It is therefore termed the pinch-off voltage and the level of "turn off" is usually some arbitrarily low value of drain current, such as 1 μA at some specified value of

Table 6-1 Definitions for the FET

TERM	DEFINITION
BV_{DGO}	Gate-to-drain breakdown voltage with source open
BV_{GSS}	Gate-to-source breakdown voltage with drain shorted to source
Channel	Semiconductor conductive path
Depletion mode	Operation of FETs in which an increase in V_{GS} results in a decrease in drain current
Enhancement mode	Operation of FETs in which an increase in V_{GS} results in an increase of drain current
Field effect	Varying the conductivity of a semiconductor by the application of a perpendicular electric field
g'_{fs}	Transconductance of an FET at I_{DSS}
I_{DS}	Drain and source current
I_{DSS}	Drain current with $V_{GS} = 0$
IGFET	Insulated-gate field-effect transistor, sometimes used as abbreviation for MOSFET
JFET	Junction field-effect transistor—conduction controlled by amount of reverse bias on a *pn* junction
MOSFET	Metal oxide semiconductor field-effect transistor—conduction controlled by insulated capacitor gate
Pinch-off voltage (V_P)	Gate-to-source voltage that causes current conduction through channel to fall to almost zero [typically 1 nanoampere (nA) to 1 μA]
Pinch-off region	Region of FET characterized by almost constant current flow for changes in V_{DS}
r_{ds}	Dynamic drain resistance seen looking into drain of an FET operating in the pinch-off region
V_{GS}	Voltage from gate to source that controls channel conduction

V_{DS}. Equation (6-7) can be differentiated with respect to V_{GS} to yield a useful expression for a JFET's transconductance, since the differential of the drain current with respect to V_{GS} is, by definition, g_{fs}, or

$$g_{fs} = \frac{\Delta I_D}{\Delta V_{GS}} = \frac{2I_{DSS}}{V_P}\left(1 - \frac{V_{GS}}{V_P}\right) \tag{6-8}$$

The transconductance of an FET with $V_{GS} = 0$ is often designated by g'_{fs}. Hence

$$g'_{fs} = g_{fs}|_{V_{GS}=0} \tag{6-9}$$

Hence from Eq. (6-8) we have

$$g'_{fs} = \frac{2I_{DSS}}{V_P} \tag{6-10}$$

6-3 MOSFET Theory of Operation

The gate of a MOSFET (sometimes called IGFET) is a very small, high-quality capacitor where the conduction through the channel is controlled by the voltage applied between the gate and the source. The MOSFET input current is thus the leakage current of the capacitor, as opposed to an input current for the JFET that is the leakage current of a reverse-biased *pn* junction. Thus the MOSFET input impedance is usually orders of magnitude greater than the JFET's (often 10^6 times as large).

A thin layer of insulating material, silicon dioxide, is deposited over the channel. Over this silicon dioxide layer is a layer of a metallic material—often aluminum. Thus the input capacitor has one metallic plate, a silicon dioxide dielectric, and a semiconductor plate. Hence the abbreviation MOS. Figure 6-5 shows the construction features of a typical MOSFET.

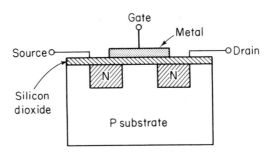

FIG. 6-5. Typical MOSFET structure

MOSFETs are available as either depletion-mode or enhancement-mode devices, depending on their construction details. Figure 6-6a illustrates the characteristics of an *n*-channel MOSFET. In the *depletion mode* the channel is normally a conductor whereby current flow may be reduced or even cut off by applying sufficient gate voltage, just as with the JFET. In the *enhancement mode* the channel is normally cut off, and may be turned on (hence *enhanced*) and controlled by application of a gate voltage. As it turns out, the depletion-mode MOSFET can also be operated in the enhancement mode, as illustrated in Fig. 6-6d.

The capacitance of the gate is very low, and therefore the input resistance very high; hence the gate is very easily charged to a voltage that might break down the capacitor's narrow silicon dioxide dielectric. Typically, a 50 V level can cause this breakdown, and this could easily be induced by the static

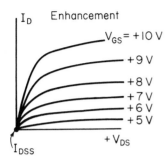
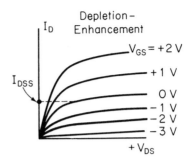

(a) Enhancement mode n-channel characteristics

(d) Depletion mode n-channel characteristics

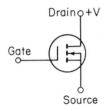
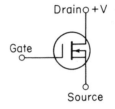

(b) n-channel enhancement mode symbol

(e) n-channel depletion mode symbol

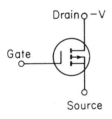
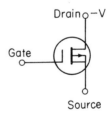

(c) p-channel enhancement mode symbol

(f) p-channel depletion mode symbol

FIG. 6-6. MOSFET characteristics and symbols

charges resulting from normal handling. For this reason these devices are often packaged with the leads shorted together. Care should be exercised in handling them by using a grounded tip soldering iron and guarding against static charges.

The frequency response of the MOSFET is better than the JFET because the inherent capacitances between gate and drain and between gate and source are generally lower for the MOSFET. Temperature effects are minimal with MOSFETs, whereas the JFET has an input leakage current that rises exponentially with temperature, as does the leakage current of any reverse-biased *pn* junction. MOSFETs are the least temperature sensitive semiconductor devices generally available today.

6-4 Biasing Techniques—Depletion-Mode FETs

There are many methods to set up an FET to prepare it for amplification of an ac signal. In this section we shall explore the most important of these methods with an eye toward the most practical circuits. For our discussion we shall consider the 2N5458 JFET. It is a general-purpose low-cost *n*-channel device, which can be purchased for less than $1.00 in small quantities and less than $0.50 in lots of 100 or more. It has the following ratings:

1. Transconductance: $g_{fs} = 1500\text{--}5500 \ \mu\mho$
2. Zero gate voltage drain current: $I_{DSS} = 2\text{--}9$ mA.
3. Leakage current: $I_{GSS} = 1$ nA max.

The setup shown in Fig. 6-7 is the simplest common-source biasing scheme. If we let $R_G = 100$ MΩ, then even at the maximum I_{GSS} (leakage

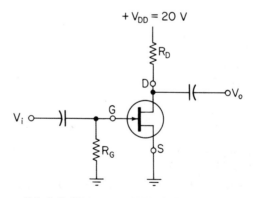

FIG. 6-7. Elementary JFET biasing scheme

current) of 1 nA, we have just a 0.1 V drop across R_G. It is therefore safe to assume that $V_{GS} = 0$ and the drain current will equal I_{DSS}. If R_D were made equal to 2 kΩ, the operating point V_D would range from

$$V_D = V_{DD} - V_{RD}$$
$$= 20 \text{ V} - 2 \text{ mA} \times 2 \text{ k}\Omega = 16 \text{ V}$$

at the minimum value of I_{DSS} of 2 mA up to

$$V_D = 20 \text{ V} - (9 \text{ mA} \times 2 \text{ k}\Omega) = 2 \text{ V}$$

at I_{DSS} of 9 mA. Obviously, this method of biasing has caused a very wide Q-point range, but this may be quite adequate for low input signal levels. However, the input signal *must* be limited to not exceed about $+0.5$ V on the gate–channel junction in order to prevent forward bias and the resulting input current flow and probable device destruction. This then implies a maximum ac input voltage of 1 V p-p. Despite these limitations, the circuit of Fig. 6-7 could practicably be put to use in the lab as a simple one-of-a-kind, low-signal-level, high-input-impedance amplifier. In addition, most FET amplifiers are used as an input stage to an amplifier system to offer a high input impedance from a high impedance signal source. This classification of signals is generally very low level—millivolts and less. Since the voltage gain of FET amplifiers is usually 10 or less, the output signal is still small enough to allow operation with the 2 to 16 V bias level range this design results in. Thus the circuit is applicable to many production designs. The voltage gain [from Eq. (6-5)] varies from

$$A_v = g_{fs}R_D = 2 \text{ k}\Omega \times 1500 \text{ }\mu\mho$$
$$= 3$$

up to

$$A_v = 2 \text{ k}\Omega \times 5500 \text{ }\mu\mho$$
$$= 11$$

through its full possible transconductance range. For its simplicity and low cost, this circuit offers an amazing impedance transformation from an input impedance of 100 MΩ down to a 2 kΩ load. The specific maximum leakage current of 10^{-9} A implies an input impedance of about 10^9 Ω, and hence we can safely assume that Z_{IN} will approximately equal R_G.

This form of bias (or really lack of bias, since $V_{GS} = 0$) is equally applicable to the *n*-channel depletion-mode MOSFET. In fact, since the MOSFET actually operates in both depletion and enhancement modes, it has the advantage of allowing larger input signals. Refer to Fig. 6-6d to verify this statement. To obtain proper operation with a *p*-channel depletion-mode MOSFET, a simple reversal of the battery polarity is required. The effects of temperature change on the MOSFET are minimal with temperature compensation seldom a requirement.

The effect of temperature on the bias point of any form of JFET amplifier is quite small as long as the gate leakage current is small enough not to cause an appreciable voltage drop (> 0.1 V) across a gate-to-ground resistor.

However, two other temperature effects can vary the bias point; fortunately, they tend to counterbalance one another. It has been determined that perfect temperature equilibrium occurs when the dc gate-to-source bias is

$$V_{GS} = V_P + 0.63 \tag{6-11a}$$

for n-channel devices and

$$V_{GS} = V_P - 0.63 \tag{6-11b}$$

for p-channel devices, where V_P is the pinch-off voltage. Recall that V_P is the gate-to-source voltage required to effectively turn off an FET's conduction. Thus in very critical situations a means is available to practically eliminate bias point drift with temperature by utilizing a very specific value of V_{GS}. It is possible to predict the amount of Q-point drift when not operating at the zero drift bias level predicted by Eq. (6-10) and (6-11). By substituting Eq. (6-10) into Eq. (6-7), we obtain the drain current at zero drift conditions, I_D'':

$$I_D'' = I_{DSS}\left(\frac{0.63}{V_P}\right)^2 \tag{6-12}$$

The following equation for Q-point drift requires the use of I_D'' predicted from Eq. (6-12):

$$Q\text{-point drift} \simeq -2.2\left(1 - \frac{I_D}{I_D''}\right) \text{mV}/^\circ\text{C} \tag{6-13}$$

for n-channel JFETS. For p-channel JFETS drop the minus sign.

EXAMPLE 6-1

A 2N5458 JFET is to be operated at zero gate bias, as shown in Fig. 6-7. Determine the amount of Q-point drift per degree Celsius if $I_{DSS} = 5$ mA and $V_P = -3$ V.

Solution:
The drain current at zero drift is [from Eq. (6-12)]

$$I_D'' = 5 \text{ mA}\left(\frac{0.63}{3}\right)^2$$

$$= 0.22 \text{ mA}$$

$$\text{drift} = -2.2\left(1 - \frac{5 \text{ mA}}{0.22 \text{ mA}}\right) \text{mV}/^\circ\text{C}$$

$$= -2.2(1 - 22.7) \text{ mV}/^\circ\text{C} = -2.2(-21.7) \text{ mV}/^\circ\text{C}$$

$$= +47 \text{ mV}/^\circ\text{C}$$

EXAMPLE 6-2

Determine g_{fs} when the JFET of Example 6-1 is operated at zero bias.

Solution:
Recall that the transconductance at zero bias is g'_{fs}. From Eq. (6-10)

$$g'_{fs} = \frac{2I_{DSS}}{V_P}$$

$$= \frac{2 \times 5 \text{ mA}}{3 \text{ V}} = 3333 \ \mu\mho$$

Figure 6-8 illustrates a bias method that also provides negative feedback. We shall concern ourselves with the biasing now and leave the feedback

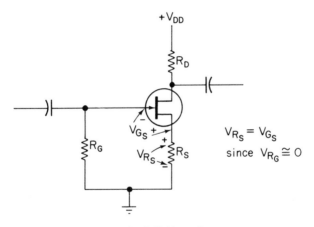

FIG. 6-8. Self-bias scheme

discussion for Section 6-6. Whatever value of I_D that results from the biasing will also flow through R_S. This will cause a dc voltage drop of the polarity shown across R_S. Using Kirchhoff's voltage law around the loop that includes R_S, R_G, and the gate–source junction, the gate–source voltage is equal to V_{RS} and is of the polarity shown. This is true if the gate leakage current is low enough to cause only a minimal voltage drop across R_G. This is normally the case; hence this is an adequate biasing scheme, and $V_{GS} = V_{RS}$.

EXAMPLE 6-3

The characteristic curves for a particular 2N5458 JFET are provided in Fig. 6-9. With a V_{DD} of 20 V and $R_D = 3 \text{ k}\Omega$, determine the necessary value of R_S to provide a 10 V Q-point at the drain.

126 Field-Effect-Transistor Amplifiers

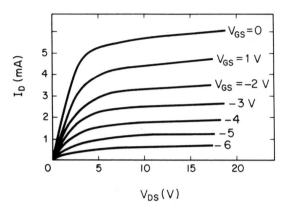

FIG. 6-9. 2N5458 characteristics

Solution:

I_D can be calculated as $3\frac{1}{3}$ mA. This corresponds to $V_{GS} = -2$ V from the characteristic curves. Therefore, V_{RS} must equal 2 V and will have $3\frac{1}{3}$ mA flowing through it. Hence

$$R_S = \frac{2 \text{ V}}{3\frac{1}{3} \text{ mA}} = 600 \, \Omega$$

The solution to Example 6-3 is unique to the 2N5458 represented by the characteristics in Fig. 6-9. Unfortunately, these curves change appreciably with changes in g_{fs}, and g_{fs} is a widely varying parameter for different 2N5458s. We are therefore confronted with a biasing scheme that is ideal for one-of-a-kind designs, but possibly unsuitable for a circuit to be constructed in high volume. A careful analysis would be necessary to ensure that the Q-point range was not excessive for all possible device-to-device parameters affecting this Q-point. In any event, the voltage gain has now been stabilized in a fashion similar to the addition of an emitter resistor to the CE amplifier because of negative feedback. A source bypass capacitor could be used to increase the voltage gain just with the CE amplifier. This form of bias is also applicable to the n-channel depletion-mode MOSFETs. Reversing the supply voltage makes it suitable for p-channel depletion FETs.

A circuit that provides a bias which is relatively independent of parameter variations is shown in Fig. 6-10. Unfortunately, it requires the use of two separate dc supplies, but many electronic systems provide that capability anyway. This method is also applicable to the depletion-mode MOSFET. A bias voltage of $+9$ V will be the design goal and ± 18 V supplies are available. If the source is to be at $+2$ V, a bias voltage of $V_{GS} = -2$ V will result, since the gate is at 0 V (because of no appreciable current flow through R_G). From

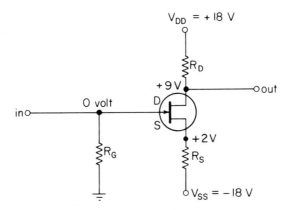

FIG. 6-10. Biasing scheme independent of parameter variation

the characteristics of the 2N5458 shown in Fig. 6-9, I_D is 3 mA when $V_{DS} = 7$ V and $V_{GS} = -2$ V. Therefore, R_D must equal (18 V − 9 V)/3 mA or 3 kΩ. The drop across R_S is 20 V, so that R_D must equal 20 V/3 mA = 6.67 kΩ.

This is the proper setup for a 9 V bias, but what happens when a different 2N5458 is used? If the new 2N5458 results in $V_{GS} = -3$ V, the drop across R_S will increase from 20 to 21 V. The drain current is now forced up to 21 V/6.67 kΩ, or 3.15 mA. Therefore, the drop across R_D will now be 3.15 mA × 3 kΩ = 9.45 V, and a change in bias point from 9 to 8.55 V has occurred. Thus a fairly drastic change in device parameters has resulted in a modest change in bias point. This biasing method is useful for all but the most demanding requirements. However, it results in voltage gains less than 1 because of the large amount of feedback. The feedback factor (B) is $-R_S/R_D$ making $A'_V = g_{FS}R_D/(1 + g_{FS}R_S)$.

EXAMPLE 6-4

Design a circuit similar to the one shown in Fig. 6-10 with ±14-V supplies and a design goal of a +7 V bias with a nominal $V_{GS} = -2$ V. Use the 2N5458 device predicted by the curves of Fig. 6-9. Determine the percentage of change in bias point that results from the use of a different JFET which gives a $V_{GS} = -3$ V in this circuit.

Solution:
The desired Q-point is +7 V at the drain, and with $V_{GS} = -2$ V the source voltage would be +2 V. The drain current from Fig. 6-9 for these conditions is about 3 mA; therefore,

$$R_D = \frac{14 \text{ V} - 7 \text{ V}}{3 \text{ mA}} = 2.33 \text{ k}\Omega$$

The drop across R_S will be 16 V; hence

$$R_S = \frac{16 \text{ V}}{3 \text{ mA}} = 5.33 \text{ k}\Omega$$

The new 2N5458 that results in a V_{GS} of -3 V when placed in this circuit causes the voltage across the source resistor to increase from 16 to 17 V. Therefore, I_D increases to

$$I_D = \frac{17 \text{ V}}{5.33 \text{ k}\Omega} = 3.18 \text{ mA}$$

and V_{R_D} will rise to

$$\begin{aligned} V_{R_D} &= V_{DD} - V_D \\ &= 14 \text{ V} - (3.18 \text{ mA})(2.33 \text{ k}\Omega) \\ &= 6.6 \text{ V} \end{aligned}$$

This is a percentage change of

$$\frac{7 \text{ V} - 6.6 \text{ V}}{7 \text{ V}} \times 100\% = 5.7\%$$

Note that the lower supply voltages tend to increase the percentage of bias change.

6-5 Biasing—Enhancement-Mode MOSFETs

The bias simplicity for the enhancement-mode MOSFET provides some interesting applications. Figure 6-11 shows some typical characteristic curves and a likely biasing scheme. Notice that the bias voltage (V_{GS}) is of the same magnitude as the Q-point voltage V_D. This is because the gate of a MOSFET draws almost no current (10^{-14} A is typical) through R_G, resulting in zero potential difference between gate and drain.

The minimal bias circuitry makes MOSFETs ideal for integrated circuits, since MOSFETs turn out to be cheaper than resistors due to their much smaller size. In fact, MOSFETs are often used as resistors in integrated circuits. We shall discuss these aspects in greater detail in Chapter 11.

Temperature effects of MOSFET bias levels are minimal, which means that the circuit of Fig. 6-11b is commercially suitable for volume production if changes in gain with g_{fs} variations are allowable. The gain of the circuit is predicted by Eq. (6-5),

$$A_v \simeq g_{fs} R_D |_{r_{ds} \gg R_D} \tag{6-5}$$

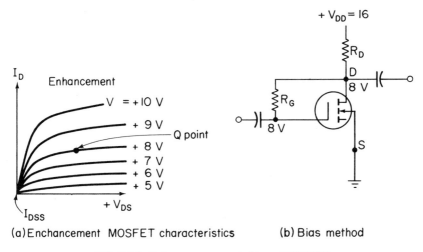

(a) Enchancement MOSFET characteristics   (b) Bias method

FIG. 6-11. Enhancement-mode bias of MOSFET

and hence can vary through a range equal to the range of g_{fs}, typically 3:1. If this is not acceptable, some means of negative feedback will solve the problem.

These devices lend themselves to many special applications—some of which we shall refer to in remaining sections of this book. The fact that the gate and drain can operate at the same potential allows for an extremely simple, direct-coupled, multistage amplifier. Figure 6-12 shows a possible

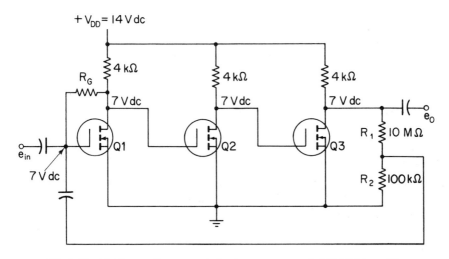

FIG. 6-12. Multistage direct-coupled enhancement-mode MOSFET amplifier

configuration. Notice that the output is fed back to the input through the $R1$, $R2$ voltage divider to provide negative feedback.

EXAMPLE 6-5

For the dc amplifier shown in Fig. 6-12, determine the overall voltage gain.

Solution:

$$G_v = g_{fs}(R_D \parallel R_L) \tag{6-6}$$

R_L is much higher than R_D for all three MOSFETs. Hence the gain of each stage can vary from

$$G_v = 2000 \; \mu\mho \times 4 \; k\Omega = 8$$

up to

$$G_v = 6000 \; \mu\mho \times 4 \; k\Omega = 24$$

Hence the overall gain is 8^3 to 24^3, or 512 to 13,824. Obviously this range of gains is intolerable for virtually any application, and negative feedback is a necessity. The feedback factor B is

$$-\frac{R_1}{R_1 + R_2} = -\frac{100 \; k\Omega}{10 \; M\Omega + 100 \; k\Omega} \cong -\frac{1}{100}$$

Therefore,

$$G'_{v \; min} = \frac{G_v}{1 - BG_v} = \frac{512}{1 + (512/100)} = 84$$

and

$$G'_{v \; max} = \frac{13{,}824}{1 + (13{,}824/100)} \cong 100$$

The amplifier is therefore quite stable (gain of 84 to 100) and very simple.

6-6 Negative Feedback Considerations

We have thus far considered some of the most likely bias setups for the various types of FETs. Once biased, the device should be ready to successfully amplify the input signal. It has been shown that because of the wide ranges of g_{fs} for a given FET, it is generally necessary to introduce negative feedback to maintain a predictable level of voltage gain. Example 6-5 showed a method of introducing this feedback (voltage divider feedback) and quite graphically illustrated its extreme power to make an amplifier's gain independent of the active device parameters. In Example 6-5 the feedback was accomplished over a number of stages, generally termed *overall feedback*. This is a common practice, but individual-stage feedback is also common, especially if bias-point temperature stabilization is also a desired result of the feedback.

The source resistor self-bias circuit already discussed (Fig. 6-8) incorporates negative feedback in the same way that an emitter resistor in a CE amplifier does. The emitter resistor does not have as direct an effect on biasing as does the source resistor, however. Just as in the CE amplifier, the feedback factor B is given by a resistor ratio. The value of B for the self-bias source resistor circuit is $-R_S/R_D$. If an external ac load is present, B is equal to $-R_S/(R_D//R_L)$.

EXAMPLE 6-6

The circuit shown in Fig. 6-13 is utilizing an FET with $g_{fs} = 4000 \ \mu\mho$. Determine G_v and G_v'.

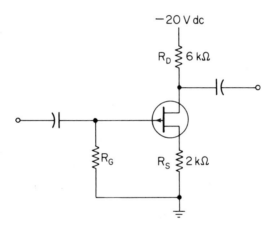

FIG. 6-13. Example 6-6

Solution:
From Eq. (6-5)

$$G_v = g_{fs} R_D \quad (6\text{-}5)$$
$$= 4000 \ \mu\mho \times 6 \ k\Omega = 24$$

The feedback factor B is $-R_S/R_D = -\frac{1}{3}$. Then from the general feedback equation,

$$G_v' = \frac{G_v}{1 - BG_v} = \frac{24}{1 - (1/3)(-24)} = \frac{24}{1 + 8} \quad (5\text{-}4)$$
$$= 2.66$$

It is seen then that the source resistor is not only a means of biasing FETs (depletion mode), but introduces negative feedback as well. If, in the circuit above, a new FET with $g_{fs} = 3000\ \mu\mho$ were used, G_v would drop from 24 to 18, but G'_v would drop from 2.66 to only 2.57, and thus a voltage gain relatively independent of g_{fs} variations is achieved. The voltage gains obtained in this example are typical of single-stage FET amplifiers with feedback. In fact, it is often necessary to operate with voltage gains of around 1 to maintain a predictable voltage gain in production circuits.

Another form of feedback that is possible is shown in Fig. 6-14. This is

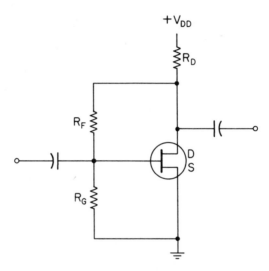

FIG. 6-14. Field-effect transistor amplifier with shunt input feedback

known as shunt input feedback, and its analysis is very similar to the CE circuit discussed in Section 5-8. Unfortunately, the input impedance is reduced $[\simeq (R_F/G_v)]$ and thus limits the usefulness of this circuit, since a high input impedance is one of the major reasons for using FETs.

6-7 Source Follower

We have thus far been concerned with common-source FET amplifiers in which the output is taken at the drain. Based upon our junction transistor experience, this would be termed a *Common Source* amplifier. One other form of FET amplifier is in common usage and it has its output at the source. It is generally termed the *source-follower* or *common-drain* amplifier. It is roughly analogus to the emitter-follower amplifier as illustrated in Fig. 6-15. Notice that although the JFET in Fig. 6-15b is a depletion mode device, some

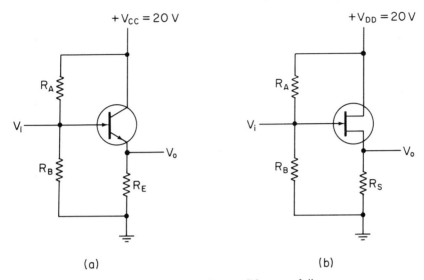

FIG. 6-15. (a) emitter follower; (b) source follower

forward bias via the voltage divider R_A and R_B is required to keep the self-bias affect of R_S from providing an excessively large negative value of V_{GS}.

It would be typical to bias the source at about one-half the supply voltage, which in the case of Fig. 6-15b would be 10 V. If no forward bias were applied, V_{GS} would be -10 V, which would virtually pinch off any FET. Since we can ignore the shunting effect of the gate on the voltage divider R_A and R_B, all that need be done is to make $R_A = R_B$ to put the gate at 10 V and leave a resulting bias of $V_{GS} = 0$.

Since all of the output signal is fed back in series with the input, the feedback factor B is -1. Hence the voltage gain of this circuit can be determined as

$$A'_v = \frac{A_v}{1 - BA_v} = \frac{g_{fs}R_S}{1 + g_{fs}R_S} \qquad (6\text{-}14)$$

where A'_v is the voltage gain with the feedback effects taken into account, and A_v is the non-feedback voltage gain.

The output impedance of the source follower is not what one would expect from a casual analysis because of the extreme feedback condition that exists. The output impedance of the common-source circuits previously analyzed is approximately the value of the resistor in the output, but in the case of the common-drain amplifier, the effect of the output resistor is often minimal.

To determine the output impedance, R_o, first let the source voltage change by 1 V. This changes V_{GS} by 1 V and will cause a source current change of $g_{fs}V_{GS}$ by the definition of transconductance. Thus the source current change is $g_{fs}(1)$ or just g_{fs}. Output impedance by definition is $\Delta V_o/\Delta I_o$ or $\Delta V_s/\Delta I_s$, which is just $1/g_{fs}$. The total output impedance is therefore $1/g_{fs} \parallel R_s$ or

$$R_o = \frac{R_S/g_{fs}}{R_S + (1/g_{fs})}$$

$$= \frac{R_S}{1 + g_{fs}R_S} \qquad (6\text{-}15)$$

EXAMPLE 6-7

Using a 2N5458 JFET, design a source follower referring to Fig. 6-16 that will bias it at I_{DSS} and at half the available 20 V supply voltage. Determine its voltage gain and R_o. Its Z_{IN} must be 50 MΩ or higher.

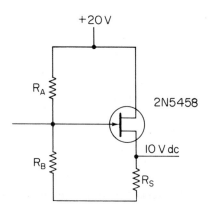

FIG. 6-16. Source-follower circuit

Solution:

In order to operate at I_{DSS} as the bias current, V_{GS} must be 0. Therefore, the gate voltage must be 10 V and R_A must equal R_B with the parallel combination of the two not to exceed 50 MΩ. Hence values of

$$R_A = R_B = 100 \text{ M}\Omega$$

are acceptable. The value of I_{DSS} varies from 2 to 9 mA, so we shall make our design for the average value of 5.5 mA:

$$R_S = \frac{10 \text{ V}}{5.5 \text{ mA}} = 1.82 \text{ k}\Omega$$

The design is now complete and the voltage gain can be calculated from Eq. (6-14):

$$A'_v = \frac{g_{fs} R_S}{1 + g_{fs} R_S} = \frac{3500 \times 10^{-6} \, \mho \times 1.82 \, k\Omega}{1 + 3500 \times 10^{-6} \, \mho \times 1.82 \, k\Omega}$$
$$= 0.865$$

using the average value of g_{fs}, which may vary from 1500 to 5500 $\mu\mho$ for the 2N5458 JFET. The output impedance is obtained from Eq. (6-15):

$$R_o = \frac{1.82 \, k\Omega}{1 + 3500 \, \mu\mho(1.82 \, k\Omega)} = 245 \, \Omega$$

EXAMPLE 6-8

For the previous example, determine the possible range for the Q-point, voltage gain, and R_o.

Solution:
With $R_S = 1.82 \, k\Omega$, the voltage gain at $g_{fs} = 1500 \, \mu\mho$ is

$$A'_v = \frac{1500 \, \mu\mho \times 1.82 \, k\Omega}{1 + 1500 \, \mu\mho \times 1.82 \, k\Omega} = 0.73$$

and at $g_{fs} = 5500 \, \mu\mho$

$$A'_v = \frac{5500 \, \mu\mho \times 1.82 \, k\Omega}{1 + 5500 \, \mu\mho \times 1.82 \, k\Omega} = 0.91$$

and hence an approximate 20 per cent range of voltage gains is expected. This is acceptable for most purposes. The R_o will vary from

$$R_o = \frac{1.82 \, k\Omega}{1 + 5500 \, \mu\mho(1.82 \, k\Omega)} = 166 \, \Omega$$

up to

$$R_o = \frac{1.82 \, k\Omega}{1 + 1500 \, \mu\mho(1.82 \, k\Omega)} = 490 \, \Omega$$

The change in Q-point becomes difficuilt to analyze, but the net result is that V_{GS} goes slightly positive for an $I_{DSS} = 2$ mA FET and the input signal must therefore be limited to perhaps several hundred millivolts peak-to-peak to keep from forward biasing the gate junction. The dc source current then is just slightly larger than I_{DSS}, and the Q-point voltage is slightly larger than $1.82 \, k\Omega \times 7$ mA or about 4 V. An $I_{DSS} = 9$ mA unit will initially cause a larger reverse bias and hence limit current until an equilibrium condition is reached. A mathematical solution is beyond the scope of this text but a good estimate would be a 12 to 15 V bias point.

The previous examples show a useful source-follower biasing scheme, but Q-point analysis becomes sticky under varying conditions. If zero V_{GS} bias is acceptable, the circuit of Fig. 6-17 performs admirably. Since the gate current is usually negligible, the voltage drop across R_G is small, resulting in the gate being at the source potential and hence $V_{GS} = 0$. The input impedance is larger than R_G, and the value of R_G can be 100 MΩ typically before its gate-leakage-current-caused voltage drop ruins the $V_{GS} = 0$ condition.

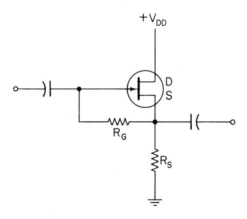

FIG. 6-17. Zero-bias source follower

6-8 Conclusions

Although all types of FETs have been introduced, all possible circuits in which they are used certainly have not been presented. The most common and elementary practical circuits have been presented, and a thorough understanding of them should enable the reader to analyze the others. It should be noted that we have concerned ourselves with FETs as linear amplifier devices, but the application of these devices in other areas is great. They are commonly used as *choppers* or *inverters* (conversion of a dc signal to alternating current to enable easier amplification), variable resistors, and switches. The use of enhancement-mode MOSFETs as switches in digital integrated circuits is enormous—far outnumbering the use of other FETs or bipolar transistors. This is due to the high physical density possible and the resulting economy of these MOSFETs in integrated circuits.

In linear amplifier applications the junction transistor is still the most often used device due to lower cost and performance considerations—mainly gain. The FET is more useful, however, when high input impedances are required, and this usually means a very small signal is involved, making zero-bias operation quite feasible. An input–output square-law relationship makes the FET a lower distortion device than others in the front end of FM radio receivers. The 3/2 power relationship of the vacuum tube or the diode-type characteristic of the bipolar transistor introduces appreciably more distortion in this critical application. The FET, on the other hand, with the square-law relation between drain current and V_{GS} [see Eq. (6-7)] makes it easier to "tune out" unwanted frequencies from the tuner's front end. It is interesting to note that at the high frequencies involved ($\cong$ 100 MHz) the input impedance of the FET is lowered to an extent that it offers no particular advantage over the bipolar transistor in that respect. This is caused by "shorting-out" effects due to inherent capacitance within the FET. We shall study these effects in Chapter 8.

Another difficulty with the use of FETs in linear amplifiers is that to maintain an acceptable level of distortion they must be operated at small-signal levels. To illustrate this point, refer to Fig. 6-18, which shows the drain

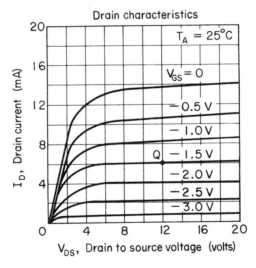

FIG. 6-18. 2N5163 drain characteristics (Courtesy of Fairchild Semiconductor.)

characteristics for a 2N5163 JFET. Large-signal operation might typically mean a 3 V p-p input applied around the bias point labeled Q on the $V_{GS} = -1.5$ V curve. Notice the greater spacing between the different curves as $V_{GS} = 0$ is approached. Referring to Fig. 6-18, the drain current shifts from about 6 mA up to about 14 mA as the input voltage goes from 0 to $+1.5$ V. If the output were at the drain and a 1 kΩ drain resistor were used, the output voltage would swing from a total of (14 mA $-$ 6 mA) $\times$ 1 kΩ or 8 V. The $1\frac{1}{2}$ V negative half-cycle input signal causes a drain current variation of from 6 mA down to about 1 mA, which results in an output voltage swing of

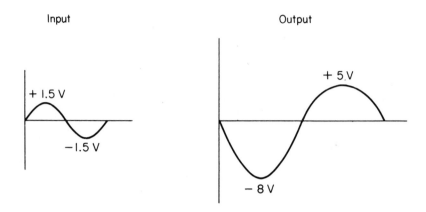

FIG. 6-19. Large-signal distortion by FET amplifier

(6 mA − 1 mA) × 1 kΩ or 5 V. The resultant situation is illustrated in Fig. 6-19. It is the positive output that gets compressed because of the signal phase reversal from gate to drain. These effects can be minimized by negative feedback techniques, but the better input–output linearity of bipolar transistors makes them much more appropriate for the task. Operation at small signals in the millivolt region with FETs is, however, usually very acceptable, which means that zero-bias operation is usually acceptable—it being the easiest attainable bias condition.

Large power applications are currently not yet possible, because power FETs are still in the developmental stage. Units capable of 10 W dissipation should be commercially available in the near future, but much work remains to be done in this area.

PROBLEMS

1. Identify and discuss the three areas of operation for a JFET.
2. Why is the dynamic drain resistance usually a very high value?
3. A JFET is operating at a drain current of 2 mA and $V_{GS} = -2$ V. If it has a pinch-off voltage of -5 V and $r_{ds} = 100$ kΩ, determine I_{DSS}, g_{fs}, and g'_{fs}.
4. If the FET in Problem 3 is operating with a 5 kΩ drain resistor (R_D), determine the voltage gain.
5. Why is a depletion-mode MOSFET also able to operate enhancement mode, whereas its JFET counterpart cannot?
6. The FET in Fig. P6-6 has $V_P = 5$ V and $I_{DSS} = 10$ mA. Determine the

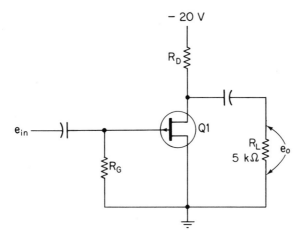

FIG. P6-6.

value of R_D to provide a -10 V bias level, and determine the voltage gain with and without the 5 kΩ ac load resistor.

7. If I_{DSS} can vary 5–15 mA in Problem 6, determine the resulting range of Q-points and voltage gains. Assume V_P remains constant.

8. Calculate the value of V_{GS} that would provide zero bias level drift for the circuit of Fig. P6-6. Determine the amount of Q-point drift per degree Celsius that would result for the circuit as designed in Problem 6.

9. Calculate the input and output impedances of the circuit designed for Problem 6. Why is R_G included in circuits of this type; i.e., what purposes does it serve?

10. Design the circuit of Fig. P6-10 to provide a $+10$ V bias at the drain using the characteristics provided by Fig. 6-9.

11. Determine the voltage gain of the previous circuit. How could its voltage gain be easily increased without affecting the Q-point, and what would that gain be?

12. Design the circuit of Fig. P6-12 to provide a $+10$ V Q-point with a value of -1 V for V_{GS}. Use the FET having the characteristics illustrated in Fig. 6-9. Once designed, a different FET is used that results in a V_{GS} of -2 V. Determine the change in Q-point.

13. Determine A_V and A_V' for Problem 12 if $g_{fs} = 3000$ $\mu\mho$. Determine the range of A_V and A_V' if g_{fs} could range from 1500 to 4500 $\mu\mho$.

14. Why is zero bias ($V_{GS} = 0$) usually an acceptable operating point with JFETs used in linear amplifiers?

15. Using a 2N5163 JFET with the characteristics shown in Fig. 6-18, design a source-follower circuit. The battery available is 12 V, a 6 V Q-point is

desired, and zero gate-to-source bias is to be used. Calculate the appropriate value of R_S, and the biasing resistor (s), the voltage gain, and the output impedance.

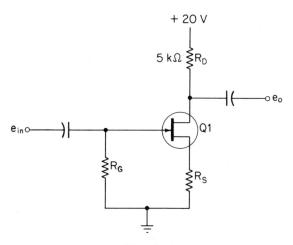

FIG. P6-10.

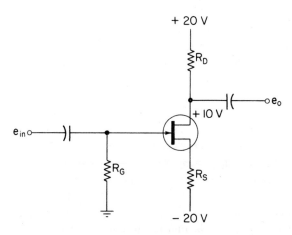

FIG. P6-12.

7

Power Amplifiers

7-1 Introduction

The amplifiers we have thus far considered were generally of the *small-signal* variety. A broad definition of a *small-signal amplifier* is that the ac input and output voltages and currents are small with respect to the quiescent dc voltage and current levels. In a *large-signal* amplifier, either the ac voltage or current (or both) levels are approaching their maximum possible value. That maximum level is determined either by the power supply employed or the transistor voltage and/or current ratings. In cases where appreciable amounts of power are involved (for instance 1 W or more), the amplifier is termed a *power amplifier*. Thus not all large-signal amplifiers are necessarily power amplifiers, but it is safe to say that most are.

Power amplifiers (PAs) are similar to those amplifiers previously studied in many ways. For instance, a schematic diagram of a single-stage small-signal amplifier (SSA) and a power amplifier might appear identical. In other words, a bias setup is required and an emitter feedback resistor would probably be present. Despite the similarities, many new aspects must be considered when dealing with PAs. It is these aspects that form the basis for this chapter.

It is the function of an amplifier to drive a load. The type of load being driven determines the required level of voltage and current. If the load is a high fidelity speaker with an 8 Ω impedance and 50 W of drive is necessary for full volume, certain requirements of the amplifier may be determined. The root-mean-square output voltage is calculated from the given values of P and R as

$$V^2 = P \times R$$
$$V^2 = 50 \text{ W} \times 8 \text{ Ω} = 400$$
$$V = 20 \text{ V rms}$$
$$V = 20 \times 2 \times \sqrt{2} \simeq 56 \text{ V p-p}$$

The ac output current is calculated as

$$I = \frac{V}{R} = \frac{56 \text{ V p-p}}{8 \text{ Ω}}$$
$$I = 7 \text{ A p-p}$$

Obviously, this is a power amplifier, and the load being driven effectively dictates the power output requirement.

If the load is a cathode-ray tube (CRT) that requires a 200 V sawtooth for horizontal deflection, one might be tempted to say that it requires a PA for drive. However, since virtually no current is drawn by the CRT, the power level is very low, requiring a large-signal amplifier but not a PA.

Since the cost of transistors is roughly proportional to their power rating, it is customary to use power transistors at close to their rated power level. For instance, it would be uneconomical to use the transistor with the characteristics shown in Fig. 7-1 in a lower power application since a lower-cost smaller device could be found to do the job. Note the dashed line indicating the maximum power dissipation level of 10 W in Fig. 7-1. Any area to the right or above that line ($V_{CE} \times I_C$) exceeds the device's power rating. Power transistors have a maximum permissible junction temperature (T_j) as do all other semiconductors. The power the transistor dissipates ($\cong I_C V_{CE}$) determines the amount of temperature rise above the temperature of the transistor's environment (ambient temperature). Operation at junction temperatures above about 110°C for germanium or 175°C for silicon devices results in device destruction. The amount of device dissipation is determined by the product of I_C and V_{CE} at all times. With a 10 W limit this describes the hyperbolic dashed line shown in Fig. 7-1.

Of special interest in Fig. 7-1 is the high voltage rating of the particular device shown. Note the 300 V maximum rating for V_{CE}. This value is close to present power transistor state-of-the-art values for maximum voltage levels. Power transistors with allowable collector currents of about 250 A are at the upper limit with respect to current. Thus there are three basic limitations on

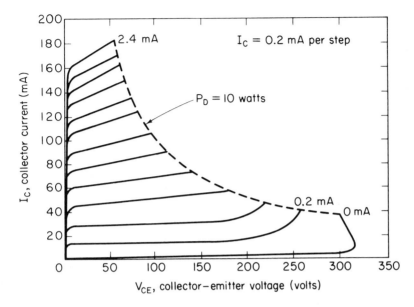

FIG. 7-1. High voltage power transistor characteristics

power transistors: voltage, current, and power. Exceeding any one of these ratings would probably result in device destruction.

Since it is normal and economical to operate a power transistor close to its absolute limits, distortion of the output signal becomes an important consideration. At both very high and very low values of collector current, the transistor's h_{fe} is reduced. Operation in these nonlinear areas results in distortion, which is either tolerated or compensated for by using negative feedback. Figure 7-1 also illustrates this effect by showing a smaller spacing between the $I_B = 0.2$ mA per step characteristics at *both* high and low collector currents. In addition, transistors exhibit a nonlinear input characteristic, which can also introduce distortion.

7-2 Heat Dissipation

Since power transistors must be able to dissipate large amounts of power, their construction must provide for efficient dissipation of heat. The greatest fraction of power dissipation takes place in the reverse-biased collector–base junction (the base–emitter junction conducts about the same amount of current but has a much lower voltage drop). It is customary to therefore have the collector electrically and physically connected to the metal transistor case.

This enables the generated heat to be efficiently conducted away from the collector (to a heat sink) and then into surrounding air or free space. The larger the heat-dissipating surface area of both case and sink, the better the heat-dissipating effect; hence the larger size of power transistors compared to small-signal devices.

In general, a power transistor must be mounted to a larger metallic surface termed a *heat sink* to maintain satisfactory temperature operation. The heat sink provides an increased surface area for heat dissipation, allowing the transistor to operate at higher power levels while still keeping the junction temperature (T_J) at an acceptable level. This is sometimes accomplished by physically mounting the transistor case directly to the chassis of the equipment. Care must then be exercised, since the collector is electrically connected to the chassis (which is usually at ground potential) unless insulated in some fashion. Mica washers are usually used for such collector isolation at low collector voltages, since they offer good thermal conductivity but very high electrical resistance. At higher voltages, special heat sinks are used instead of just attaching the transistor to the chassis. These are usually made of aluminum, which offers excellent thermal conductivity with low weight. The aluminum is often extruded or corrugated to offer the maximum surface area possible and may also be painted black to maximize its heat-dissipating ability.

Thermal resistance (θ) is the relationship that expresses the temperature rise (ΔT) per watt (P) of dissipation power or

$$\theta = \frac{\Delta T}{P} \tag{7-1}$$

The thermal resistance from the transistor *junction* to *case* is denoted by θ_{JC}, where

$$\theta_{JC} = \frac{T_J - T_C}{P} \tag{7-2}$$

T_J and T_C are temperatures of the junction and case, respectively, and the thermal resistance from the transistor junction to ambient is given by θ_{JA}:

$$\theta_{JA} = \frac{T_J - T_A}{P} \tag{7-3}$$

The values of these parameters are usually given in power transistor specification sheets and are used to ensure proper thermal design and to calculate required heat sink areas, as given in Eq. (7-5).

There is also a small thermal resistance between the transistor case and the heat sink, which is θ_{CS} or

$$\theta_{CS} = \frac{T_C - T_{\text{sink}}}{P} \tag{7-4}$$

The *surface area* of the heat sink determines its thermal resistance to the surrounding environment (ambient) or

$$\theta_{SA} = \frac{T_{\text{sink}} - T_{\text{ambient}}}{P} \tag{7-5}$$

Figure 7-2 is a graph of heat-sink area versus θ_{SA} for flat square sheets of $\frac{1}{8}$ in. unpainted aluminum mounted vertically. The minimum acceptable

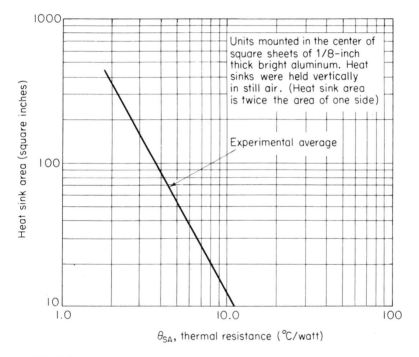

FIG. 7-2. Heat sink area versus thermal resistance (Courtesy of Motorola Semiconductor Products, Inc.)

value for θ_{SA} is given by the following relationship:

$$\theta_{SA} = \frac{T_{J\,\text{max}} - T_{A\,\text{max}}}{P_{\text{max}}} - \theta_{CS} - \theta_{JC} \tag{7-6}$$

where all terms are as defined previously. The thermal resistance from case to sink, θ_{CS} in Eq. (7-4), depends on the amount of contact between the two surfaces. Its value (θ_{CS}) is decreased (thus improving temperature conduction) by the use of a silicone grease compound, which serves to maximize the surface contact. Typical values of θ_{CS} are 0.2°C/W with no electrical insulator and 0.8°C/W with a mica insulator. These values are cut in half when a silicone grease is used.

The power rating of a power transistor is only valid when the transistor-case temperature is held to 25°C ($\simeq$ room temperature). This temperature is difficult to maintain even with a heat sink, and it therefore is necessary to derate the transistor according to the expected maximum case temperature, and *derating* information is supplied by the transistor manufacturer. Figure 7-3 shows the power-versus-temperature *derating curve* for a medium-power

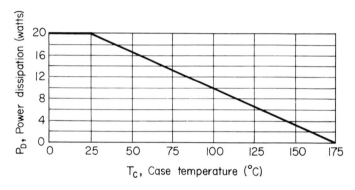

FIG. 7-3. Power derating curve for 2N3766 transistor (Courtesy of Motorola Semiconductor Products, Inc.)

2N3766 transistor. It has a 20 W rating, but if used above 25°C case temperature, it must be derated 0.133 W/°C. Notice that the power rating falls to 0 W at 175°C, since this is the maximum allowable junction temperature. Any power dissipation at 175°C would raise the junction temperature above this maximum of 175°C. It is presumed that a junction temperature above 175°C causes device failure. The maximum case temperature of a power transistor can be calculated, if its junction-to-case thermal resistance (θ_{JC}) is known, by the following formula:

$$T_{\text{case max}} = T_{\text{ambient max}} + \theta_{JC} \times P_D \tag{7-7}$$

EXAMPLE 7-1

A 2N3766 transistor is used without a heat sink in an application in which the maximum continuous power dissipation is 10 W and the maximum ambient temperature reaches 50°C. This device has a θ_{JC} rating of 7.5°C/W. Is this operation within the rated limits of this transistor?

Solution:

$$\begin{aligned} T_{\text{case max}} &= T_{\text{ambient max}} + \theta_{JC} \times P_D \\ &= 50°C + 7.5°C/W \times 10\ W \\ &= 50°C + 75°C \\ &= 125°C \end{aligned} \quad (7\text{-}7)$$

Now from Fig. 7-3 we find a maximum power rating of only 6.5 W at a 125°C ambient operation. On the other hand, if the ambient temperature never rises above 25°C, the maximum case temperature would be 100°C. This corresponds to a 10 W allowable dissipation; just acceptable.

EXAMPLE 7-2

Determine the required heat-sink size to allow acceptable operation for Example 7-1. The transistor is to be mounted to the aluminum using no silicone grease and a mica washer.

Solution:

$$\begin{aligned} \theta_{SA} &= \frac{T_{J\ \text{max}} - T_{A\ \text{max}}}{P_{\text{max}}} - \theta_{CS} - \theta_{JC} \\ &= \frac{175°C - 50°C}{10\ W} - 0.80°C/W - 7.5°C/W \\ &= 12.5°C/W - 0.80°C/W - 7.5°C/W \\ &= 4.2°C/W \end{aligned}$$

Now from the graph of Fig. 7-2 we determine a heat-sink area requirement of 80 in.^{2}, corresponding to a θ_{SA} requirement of 4.2°C/W. Thus a plate 8 × 5 in. would satisfy the requirement, remembering that *both* sides of the plate are considered for total surface area (see note on Fig. 7-2).

Heat-sink calculations such as the foregoing should be considered as *approximations*. The many factors that determine the ability of a heat sink to transfer heat into the ambient space make this precaution vital. The heat sink should be operated under the worst possible conditions, and the case temperature monitored to ensure adequate performance. In many severe applications, fans (or even circulating water) may be used to remove heat buildup. It should be noted that heat-sinking techniques are also often used for resistors and diodes in medium- and high-power applications.

7-3 Amplifier Distortion Effects

As previously noted, distortion is a significant problem in power amplifiers. The following discussion, however, also applies to small-signal amplifiers, but to a much lesser degree. The amplified output signal from an *ideal amplifier* is an *exact replica* of the input signal. All practical amplifiers introduce the following types of distortion in varying degrees:

1. Amplitude distortion.
2. Frequency distortion.
3. Phase distortion.

The second and third distortions are generally secondary effects compared to amplitude distortion, but in certain situations may prove very troublesome. *Frequency* distortion occurs when certain frequencies are amplified more (or less) than others. It is caused by the capacitive and inductive effects within a circuit since the impedance of these reactances varies with frequency. We shall consider frequency distortion effects in more detail in Chapter 8.

Phase distortion results when one frequency component of a complex input signal takes a longer time to pass through an amplifier than another. Although both frequency components may be equally amplified (no frequency distortion), one has suffered a greater time delay, and hence the composite output signal may be considerably distorted.

Amplitude or *nonlinear distortion* occurs when the amplifying device is operated in a nonlinear area. All active devices have some degree of nonlinearity. This nonlinearity generates unwanted harmonics of the fundamental frequency or frequencies being amplified. It can be shown mathematically that a distorted sine wave is made up of pure sine-wave components containing the frequency (f) of the original signal and its harmonic multiples. Thus if f is known as the *fundamental* frequency, then $2f, 3f, 4f, \ldots,$ etc., are known as the *unwanted* harmonics. Often this type of distortion is characterized by a predominance of *even* harmonics or *odd* harmonics, as illustrated in Fig. 7-4. Figure 7-4a shows the result of adding just the second harmonic to a sine wave. Note that the resultant is no longer a sinusoidal wave.

Figure 7-4b shows the third harmonic added to a fundamental sine wave. These are severe cases of the distortion produced by operation in the nonlinear portions of the characteristic curve of Figs. 7-4a and b. The curves of I_C versus V_{in} are nonlinear at both low and high values and take into account that there is a nonlinear relationship between the input current (i_b) and the input voltage (V_{be}) at low levels, and a nonlinear relationship between i_b and i_c at both low and high levels.

The percentage of distortion depends upon the magnitude of the generated harmonics. The percentage of distortion D is

$$D = \sqrt{\left(\frac{V_2}{V_1}\right)^2 + \left(\frac{V_3}{V_1}\right)^2 + \left(\frac{V_4}{V_1}\right)^2 \cdots \left(\frac{V_n}{V_1}\right)^2} \times 100\% \qquad (7\text{-}8)$$

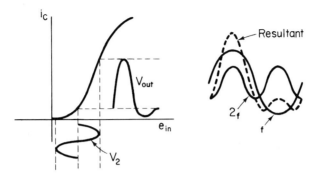

(a) "Even-harmonic" distortion — second harmonic in phase with fundamental

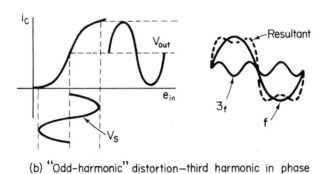

(b) "Odd-harmonic" distortion — third harmonic in phase with fundamental

FIG. 7-4. Distortion harmonic effects

where V is the root-mean-square value of voltage and the subscripts refer to multiples of the fundamental frequency. Elaborate electronic equipment is available to measure the percentage of amplitude distortion, based on an ability to "tune in" to each harmonic frequency individually and measure its amplitude. The result is presented automatically by the equipment or manually by the operator. It is uncommon for components above the fifth harmonic to have any significant effect on the amount of total amplitude distortion.

A mathematical solution for the percentage of amplitude distortion in an amplifier is possible given an accurate curve of the load current versus input voltage, as can be obtained in the laboratory. If, for example, the curve shown in Fig. 7-5 is obtained and the output current at five points is determined from the curve, the amplitudes of the fundamental and the second, third, and

150 Power Amplifiers

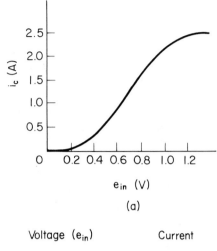

Voltage (e_{in})		Current
Positive peak 1.1 V	—	I_{max} = 2.25 A
$\frac{1}{2}$ Positive peak 0.9 V	—	I_1 = 1.8 A
At zero signal 0.7 V	—	I_Q = 1.25 A
$-\frac{1}{2}$ Positive peak 0.5 V	—	I_2 = 0.6 A
Negative peak 0.3 V	—	I_{min} = 0.15 A

(b)

FIG. 7-5. Amplitude distortion analysis: (a) amplifier transfer curve; (b) current definition for forumlas of Eq. (7-9)

fourth harmonics are determined from the following relations:

$$\left.\begin{aligned}
V_0 &= \tfrac{1}{6}(I_{max} + 2I_1 + 2I_2 + I_{min}) - I_Q \\
V_1 &= \tfrac{1}{3}(I_{max} + I_1 - I_2 - I_{min}) \\
V_2 &= \tfrac{1}{4}(I_{max} - 2I_Q + I_{min}) \\
V_3 &= \tfrac{1}{6}(I_{max} - 2I_1 + 2I_2 - I_{min}) \\
V_4 &= \tfrac{1}{12}(I_{max} - 4I_1 + 6I_Q - 4I_2 + I_{min})
\end{aligned}\right\} \qquad (7\text{-}9)$$

where the currents involved are as identified in Fig. 7-5b. If this amplifier is operated at a 0.7 V input bias level and the input signal is 0.8 V p-p, the resulting values for I_{max}, I_1, I_Q, I_2, and I_{min} are those shown. The resulting distortion calculations are then

$$\begin{aligned}
V_0 &\simeq \tfrac{1}{6}[2.25 + 2(1.8) + 2(0.6) + 0.15] - 1.25 \\
&\simeq -0.05
\end{aligned}$$

$$V_1 \simeq \tfrac{1}{3}(2.25 + 1.8 - 0.6 - 0.15)$$
$$\simeq 1.08$$
$$V_2 \simeq \tfrac{1}{4}[2.25 - 2(1.25) + 0.15]$$
$$\simeq -0.025$$
$$V_3 \simeq \tfrac{1}{6}[2.25 - 2(1.8) + 2(0.6) + 0.15]$$
$$\simeq -0.05$$
$$V_4 \simeq \tfrac{1}{12}[2.25 - 4(1.8) + 6(1.25) - 4(0.6) + 0.15]$$
$$\simeq +0.025$$

The negative sign indicates a 180° phase reversal of that component. The V_0 component is the dc component introduced by the rectification capability of the input diode (base–emitter junction). It would cause a base-to-ground Q-point shift of the magnitude calculated for V_0. Thus this distortion analysis yields an important insight into a cause for bias point shifts—stages operating with large signals will often tend to cause a Q-point shift due to rectification of the input signal.

The percentage of each harmonic can be calculated as simply the ratio of its amplitude to the fundamental (V_1) times 100 per cent. Hence the percentage of the second harmonic equals

$$\frac{V_2}{V_1} \times 100\% = \frac{0.025}{1.08} 100\% = 2.3\%$$

The total distortion as predicted by Eq. (7-8) is

$$D = \sqrt{\left(\frac{0.025}{1.08}\right)^2 + \left(\frac{0.05}{1.08}\right)^2 + \left(\frac{0.025}{1.08}\right)^2} \times 100\%$$
$$= \sqrt{0.00053 + 0.0021 + 0.0053} \times 100\%$$
$$= 0.056 \times 100\% = 5.6\%$$

Since a smaller input signal results in less distortion, it stands to reason that the addition of negative feedback might effectively reduce distortion. This is definitely the case with the loss in gain made up for with an additional stage or two, and with the overall distortion still much less than the distortion without feedback. In fact it can be shown that the distortion with feedback (D') of an amplifier is related to the nonfeedback distortion (D) as follows:

$$D' = \frac{D}{1 + BD} \tag{7-10}$$

where B is the feedback factor as defined in Eq. (5-4), and D is the percentage of distortion without feedback. The similarity between this equation and the general feedback equation of Chapter 5 is apparent.

7-4 Class A Power Amplifiers

Class A may be defined as a method of bias that allows for a complete replica of the input of an amplifier at the output. Thus the transistor neither saturates nor cuts off during the complete range of the input signal. All the amplifiers we have thus far discussed have been of this category, and therefore class A operation is nothing new to you. The class of operation becomes important in power-amplifier analysis, since deliberate operation at a transistor's cut off (class B) offers some real advantages in efficiency and economy. The last portion of this chapter will be devoted to class B operation and a variation, class AB.

The most elementary form of power amplifier is the emitter follower. Consider the amplifier in Fig. 7-6. In this amplifier the input impedance

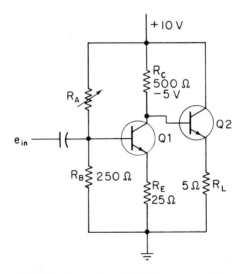

FIG. 7-6. Two-stage CE, CC power amplifier

is about 250 Ω (R2), and the load is ultimately driven to 10 V p-p into the 5 Ω load. The first stage ($Q1$) is a CE configuration with R_A and R_B used to set up the bias voltage of 5 V at $Q1$'s collector. R_E is used to introduce negative feedback and for Q-point stability. Stage 2 ($Q2$) is a CC (emitter follower) that is directly coupled to the stage 1 collector, and thus its base voltage is equal to the 5 V $Q1$ collector voltage. Its Q-point (dc voltage at the output terminal) is 5 V minus its forward-biased base–emitter voltage of about 0.7 V. Thus a 4.3 V bias is realized at $Q1$'s emitter.

The voltage gain of stage 1 should be approximately R_{L1}/R_E. The sometimes confusing part of this is that R_{L_1} is not equal to R_C, but is equal to R_C in parallel with the input impedance of stage 2, Z_{IN2}. Since h'_{fe} of $Q2$ (a CC amplifier) is given as 100, Z_{IN2} is

$$Z_{IN2} \simeq h'_{fe} R_L \simeq 500 \, \Omega$$

Thus R_{L_1}, the load on stage 1, is $500 \, \Omega \| 500 \, \Omega$ or $250 \, \Omega$. Thus the voltage gain of stage 1 is

$$G_{V1} = \frac{R_{L1}}{R_{E1}} = \frac{250 \, \Omega}{25 \, \Omega} = 10$$

If we assume ideal CC conditions, the voltage gain of stage 2, G_{V2}, is 1, and therefore the overall voltage gain G_{voa} is

$$G_{voa} = G_{V1} \times G_{V2} = 10 \times 1 = 10$$

Therefore, to provide a 10 V p-p signal to the load, V_{in} must be able to supply 1 V p-p into the 250 Ω input resistance of the amplifier.

Stage 1 is therefore a large-signal amplifier, but not a power amplifier, since it is delivering a low amount of power:

$$P = \frac{V^2}{R} = \frac{(10 \text{ V p-p}/2\sqrt{2})^2}{500 \, \Omega \| (h'_{fe} \times 5 \, \Omega)}$$
$$= 50 \text{ mW}$$

Stage 2 is a power amplifier, since it is delivering

$$P = \frac{(10 \text{ V p-p}/2\sqrt{2})^2}{5 \, \Omega}$$
$$= 2.5 \text{ W}$$

This is the ac power delivered to the load. Also of interest is the amount of dc power supplied to this circuit from the battery. With no ac input signal, the battery supplies 10 V with 5 V across 5 Ω or 5 V/5 Ω = 1 A. The current supplied will vary sinusoidally as the ac signal is being amplified between 0 V/5 Ω = 0 A and 10 V/5 Ω = 2 A at the two signal extremes. Thus the average power supplied to stage 2 is 10 V at 1 A or 10 W. The *efficiency* (η) of a power amplifier is defined as the ratio of *ac* power *out* to *dc* power *in*:

$$\eta = \frac{P_{OUT \text{ ac}}}{P_{IN \text{ dc}}} \tag{7-11}$$

As a percentage,

$$\eta = \frac{P_{OUT \text{ ac}}}{P_{IN \text{ dc}}} \times 100\% \tag{7-12}$$

Thus, for our example,

$$\eta = \frac{2.5 \text{ W}}{10 \text{ W}} \times 100\% = 25\%$$

This 25 per cent efficiency is the maximum obtainable in a class A amplifier. In reality this figure will be less, since the peak-to-peak output voltage cannot fully equal the dc supply voltage without introducing clipping. This is due to the saturation voltage of a transistor, which might typically be $V_{CE \text{ sat}} = 1$ V. Thus, even if perfect biasing existed for the remaining 9 V that are usable, a maximum 9-V p-p signal would be the absolute limit. To ensure that clipping of the output signal did not occur, it would be wise not to let the signal exceed 8 V p-p, which would mean an efficiency of

$$\eta \simeq \frac{(8 \text{ V p-p}/2\sqrt{2})^2/5 \text{ }\Omega)}{10 \text{ W}} \times 100\% = 16\%$$

This is a typical class A circuit efficiency. At the power levels being dealt with (several watts), this may be acceptable, but when dealing with higher powers (say, 10 W or more), the cost of the wasted power becomes too great to tolerate, and class B circuitry is then invariably used.

The power rating required of $Q2$ in Fig. 7-6, at no ac signal conditions, is approximately equal to $V_{CE}I_C$, which in this case averages out to 5 V $\times$ 1 A = 5 W. At full ac signal conditions the average power dissipated is unchanged; thus, a 5 W transistor is required. Many signals that an amplifier must amplify are off more than they are on—the human voice being one example. We have been considering the ideal case for class A amplifiers, which requires a transistor power rating double the ac load power. In a practical situation a good rule of thumb is to maintain a 2.5 : 1 ratio or higher.

The power amplifier thus far considered delivers both dc and ac power to the load. In fact, the ideal case delivered 2.5 W of ac power to the load at full signal and dissipated 5 W in the transistor. Since the battery was supplying 10 W of power, this leaves 2.5 W of dissipation unaccounted for. That 2.5 W is dissipated as a dc power in the 5 Ω load—a situation that often is unacceptable. A good example is a loudspeaker in which the dc current may destroy the speaker due to excessive heat or, at best, will cause poor speaker performance. A solution to this problem is to capacitively couple the load, as shown in Fig. 7-7. In this circuit, $Q2$'s emitter resistor allows a dc bias voltage to be set up, and the coupling capacitor keeps dc from entering the load. An additional feature of this circuit is an increase in efficiency, since there is no longer any dc power dissipation in the load. The maximum theoretically attainable efficiency η is now 50 per cent if $R_E = \infty$, but in

Class A Power Amplifiers 155

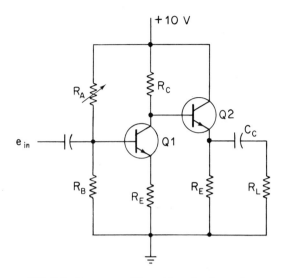

FIG. 7-7. Power amplifier with ac load coupling

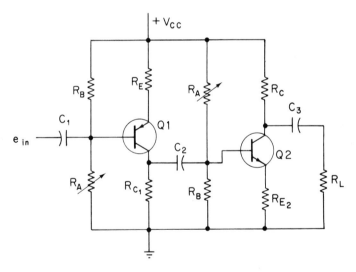

FIG. 7-8. CE, CE class A power amplifier circuit

practice 30–40 per cent is typical, since a finite (low) value of R_E is required.

Another possible class A power amplifier circuit is shown in Fig. 7-8. The major difference here is that the power stage, $Q2$, is now a CE circuit instead of an emitter follower. This allows for voltage and current gain in both stages, which in some cases might be an advantage, but several drawbacks over the previous circuits are introduced. First, notice that direct coupling between stages is no longer possible. If the base of $Q2$ were directly connected to the collector of $Q1$, the forward bias on $Q2$ would cause it to saturate. It is convenient to direct couple a CC stage to the output of a CE amplifier, but it is not normally possible in a PA to connect a CE to another CE directly. It is possible to do this in small-signal amplifiers, as we shall show in Chapter 11. The inability to directly couple means that $Q2$ must have its own bias resistors, and a coupling capacitor is necessary to isolate the dc levels of the two stages. Thus three additional passive components are necessary in Fig. 7-8 (two resistors and one capacitor), and the coupling capacitor $C2$ must be a high-valued electrolytic unit if good low-frequency response is necessary. An emitter feedback resistor R_{E2} for stability in the power stage is also required. This results in added power dissipation and thus lower overall efficiency. In general, the use of an emitter-follower circuit for the output of a class A amplifier is the logical choice over a CE stage.

One final note on the amplifier of Fig. 7-8. Since a *pnp* transistor was used for stage 1, it was necessary to "flip" it over, schematically speaking, to allow for proper voltage polarities. $Q2$ is still a CE stage, since the output is taken at the collector and the input is at the base. Various combinations of *pnp* and *npn* transistors in complementary symmetry are often utilized in multistage amplifiers, requiring careful schematic analysis to allow comprehension of the circuit.

7-5 Class A Transformer Coupling

The most common method of coupling a load to a vacuum-tube PA, until several years ago, was with a transformer. Voltage and impedance levels were thereby easily transformable, and the practice also kept direct current out of the load. Since vacuum tubes are basically high-voltage devices, transformers offered the best solution toward working into a low-voltage load. The large size and weight of transformers were not that much of a disadvantage then, since the vacuum tubes themselves were already in that category. Then as transistors began replacing tubes, a slow evolutionary process began eliminating transformers for load-coupling purposes. The transformers were beginning to be the largest and most costly elements in a PA, and circuit designers began to find alternative schemes to deliver power to a load. This evolutionary process has now reached the point where transformers are used only for very special applications—typically high voltage—when other alternatives do not exist. It is because of these special applications, and because today's electronic personnel are often required to deal with yesterday's equipment, that an introductory coverage is provided here. Transformer coupling between small-signal amplifiers (SSAs) in communications receivers is still a common practice and will be dealt with in Chapter 8.

Figure 7-9 shows the symbol for a coupling transformer. The number of turns in the primary (input side) is designated by n_p and in the secondary

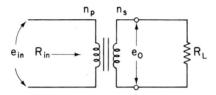

FIG. 7-9. Transformer symbol

(output) by n_s. The input voltage is related to the output voltage as follows:

$$e_o = \frac{n_s}{n_p} e_{\text{in}} \qquad (7\text{-}13)$$

The input impedance is given by

$$R_{\text{in}} = \left(\frac{n_p}{n_s}\right)^2 R_L \qquad (7\text{-}14)$$

Transformer efficiencies of 90 per cent and above are typical, which means that if 10 W of power is fed into the primary, 9 W or more can be delivered to the load.

As a practical example, consider a requirement for an amplifier that is to deliver 2.5 W at a 100 V p-p level. In many cases, a 100 V power source might not be available or economically feasible to enable a transformerless design. Another factor favoring the use of a transformer in this case is the relatively high cost of a power transistor with a breakdown voltage rating of 100 V or greater. Figure 7-10 shows a possible amplifier configuration that utilizes the basic design introduced in the previous section. It does not require a high dc supply voltage or high-voltage transistor.

Assuming ideal conditions, this circuit can deliver 10 V p-p to the transformer primary. Therefore, the turns ratio (n_s/n_p) must be 10:1 in order to deliver 100 V p-p to the output. If we assume a 100 per cent transformer efficiency to simplify calculations, then

$$R_L = \frac{V^2}{P} = \frac{(100 \text{ V p-p}/2\sqrt{2})^2}{2.5 \text{ W}}$$
$$= 5000 \ \Omega$$

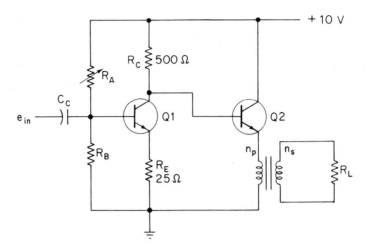

FIG. 7-10. Transformer-coupled power amplifier

Then the ac impedance presented to the amplifier is predicted by Eq. (7-14) as

$$R_{\text{IN}} = \left(\frac{n_p}{n_s}\right)^2 R_L$$
$$= \left(\frac{1}{10}\right)^2 500 \ \Omega$$
$$= 5 \ \Omega$$

Thus the ac load impedance is the same as the 5 Ω load of the circuit we analyzed in the previous section.

Since the emitter of $Q2$ is at about 5 V, there is a 5 V drop across the transformer primary. Care must be exercised to ensure that the dc winding resistance of the primary is not so low as to draw excessive dc current. In this case, a dc winding resistance of 5 Ω is typical and would result in the identical dc conditions of the transformerless circuit in Fig. 7-6. Manufacturers' data sheets should be consulted to determine this dc winding resistance.

Transformer coupling limits the low-frequency response of an amplifier stage with more core iron required to allow good low-frequency response.

7-6 Push–Pull Operation

Push–pull amplifiers operating class B make up the largest majority of linear power amplifiers. The transistors in class B amplifiers operate over one half of the input signal with the transistor cut off during the other half-cycle. Figure 7-11 shows the result of applying a sine wave to a single transistor amplifier operating class B. Amplification of the positive half only of the input signal has occurred, but rectification has also taken place. A lack of any bias on a transistor will yield this result, as shown in Fig. 7-12. During the positive half-cycle of the input signal, the transistor becomes forward

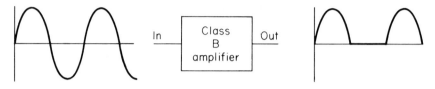

FIG. 7-11. Class B operation

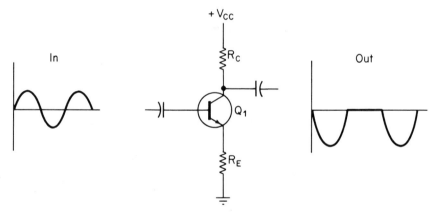

FIG. 7-12. Class B transistor amplifier

biased, and the pictured out-of-phase output signal is obtained. The negative half-cycle of the input signal reverse biases the base–emitter junction, and hence no ac output signal is obtained. This amplifier by itself then is not of much value because of the extreme distortion that it introduces into the output signal. However, if two transistors are properly connected, their combined outputs will result in a good replica of the input, as shown in Fig. 7-13. Operation of two class B transistors in this fashion is known as *push–pull operation*.

The major advantage of push–pull operation is simply a matter of more power to the load with less power dissipated in the active devices. With no ac input signal applied to the amplifier of Fig. 7-12, the only current being drawn from the dc supply is a very small leakage current. This contrasts with a class A stage, which is always biased at some significant level of quiescent current. When an input signal is applied to the class B amplifier, the current flow

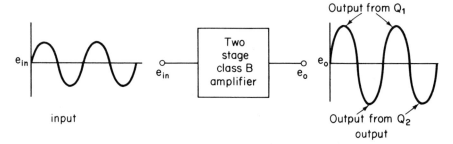

FIG. 7-13. Two-stage class B amplifier representation

through the transistor increases as the transistor voltage drop (V_{CE}) is decreasing. Thus class B transistor power dissipation is minimized, since when its V_{CE} is highest, the current flow through it is lowest (a leakage current), and when the current flow reaches a maximum, V_{CE} falls to some minimum value. The maximum theoretical efficiency of a class B push–pull amplifier is 78 per cent, at which point the power rating of each transistor must be only one tenth the ac output power. Thus, for the theoretical case, a 100 W output power requirement would dictate two 10 W transistors for class B operation or one 200 W transistor for class A operation. Not only would a massive heat sink be required for the class A design, but a more expensive, higher-current dc power source would be necessary. Another important consideration is that virtually zero power is dissipated under no-signal conditions for the class B design, but the full 200 W is being dissipated by the class A transistor under no-signal conditions. For all these reasons, virtually all PAs with over several watts output operate class B.

Recall that a practical class A design requires a transistor power rating of about 250 per cent of the output power to allow for a safety factor and the somewhat less than ideal operating conditions that all amplifiers operate in. In a similar fashion, the class B push–pull amplifier transistor ratings require padding. Instead of using transistors with 10 per cent of the required ac output power as their rating, a good rule of thumb is to use a 15–20 per cent factor.

Figure 7-14 shows two of the many possible class B push–pull circuit configurations. In Fig. 7-14a transformer coupling is utilized at the input and output. The effect of the center-tap input is to provide phase reversal of the signals applied to each base, as shown. This is necessary to allow one transistor to be "on" while the other is "off," and vice versa. If the same signal were applied to both bases, they would both be "on" and "off" at the same time–resulting in half-wave output. The output transformer then combines the two half-wave outputs into the complete signal. This type of design will be found in many older circuits but is now seldom used due to the demise of transformers.

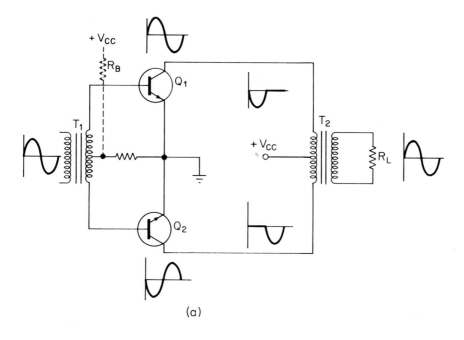

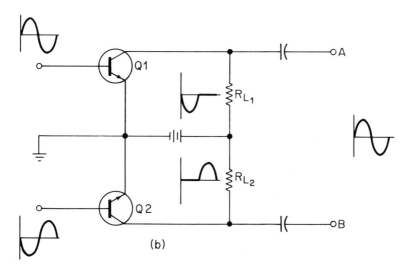

FIG. 7-14. Possible class B configurations: (a) transformer coupling; (b) transformerless coupling

162 Power Amplifiers

This same circuit (Fig. 7-14a) could easily be converted to a class A push–pull design by the resistor addition shown with dashed lines. Thus R_B would provide forward bias to each transistor through the secondary winding of $T1$. This technique is occasionally used in designs requiring extremely low distortion levels. The class A push–pull design has the inherent advantage of canceling the even-order harmonics, which does not occur for the class B design. However, the low class A efficiency is still in effect and thereby minimizes its use.

The circuit of Fig. 7-14b shows a transformerless design, if some means of generating two equal but out-of-phase input signals can be accomplished. The resulting output from A to B must be left floating with respect to ground, which is a disadvantage in some situations.

7-7 Additional Push–Pull Amplifier Considerations

The most widely used method of obtaining the out-of-phase inputs required for the circuit of Fig. 7-14b is shown in Fig. 7-15. Circuits that perform this function are termed *paraphase amplifiers, phase splitters,* or

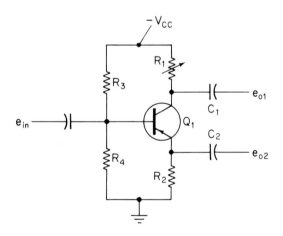

FIG. 7-15. Generation of two equal out-of-phase signals

phase inverters. The circuit shown provides the 180° phase reversal, since the collector output inverts its input signal while the emitter output does not. If $R1$ is made approximately equal to $R2$, the two outputs will be equal in magnitude, since the ac current through each is about the same ($i_e \simeq i_c$). The voltage gain will then be about 1, since $A_v = R_L/R_E = R1/R2$ and $R1 \simeq R2$. The problem with this circuit is that the two output impedances are unequal, since looking back into the emitter a low impedance is seen ($\simeq R2 \| (h_{ib} + (R3 \| R4)/h'_{fe})$), while looking back into the collector the impedance equals approximately $R1$. This difference can be compensated for by adding some resistance in series with $C2$.

An even better solution to phase-splitting problems is to use a circuit that does not require the two equal but 180° out-of-phase input signals. The use of one *npn* and one *pnp* transistor makes such a design possible. Almost all today's PAs utilize this technique. Figure 7-16 shows a possible design

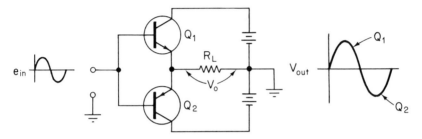

FIG. 7-16. *npn-pnp* class B push-pull amplifier

for this technique, and this circuit arrangement is known as *complementary symmetry*, since *npn–pnp* transistor combinations are called *complementary pairs*. Both $Q1$ and $Q2$ are operating as emitter followers. $Q1$ will conduct only during the positive half-cycle of the input, whereas $Q2$ is cut off by this same signal. The situation reverses itself during the input's negative excursion, with the resulting combination of signals through R_L, as shown in Fig. 7-16.

The use of identical or *matched* transistors (matched complementary pairs) will provide equal positive and negative output signals and thereby minimize distortion. This procedure is even more important in CE configurations since the feedback is not as significant as for the emitter follower, and hence the output signal is more dependent on transistor parameters. The desirability of using matched transistors is equally important for push–pull amplifiers using two *npn*s or two *pnp*s.

Unfortunately, class B operation introduces severe distortion at the very low signal levels due to the turn-on voltage of a transistor ($\simeq 0.7$ V) and the nonlinearity in the low signal area. The distortion so introduced is termed *crossover distortion* since it occurs during the time that operation is crossing over from one transistor to the other in the push–pull amplifier. Figure 7-17 provides a dynamic transfer characteristic for a class B amplifier with a slightly exaggerated amount of crossover distortion to clearly illustrate the effect. The bias point for $Q1$ and $Q2$ is at the origin, as shown.

Most applications cannot tolerate this amount of distortion, and to

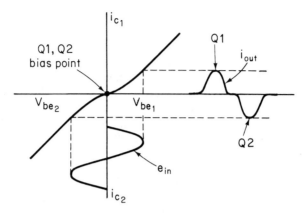

FIG. 7-17. Crossover distortion

eliminate it *class AB* operation is used. Class AB operation gives each transistor in the push–pull amplifier a small forward bias. It is defined as transistor conduction for more than 180° but less than 360°. Thus instead of operating for exactly 180° as in class B, each transistor may operate for perhaps 200° out of a full 360° cycle, which results in a small overlap or period of time when both transistors are providing output current. A typical transfer characteristic for a class AB push–pull amplifier is shown in Fig. 7-18. The dashed

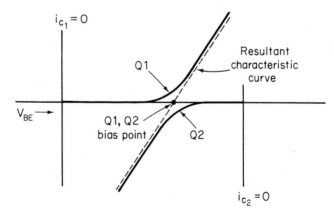

FIG. 7-18. Resultant characteristic curve for class AB operation

line shows the composite curve that is the result of both stages. The resultant is a linear relationship, as shown, which should now faithfully reproduce the input signal. The small forward bias causes the class AB amplifier to be somewhat less efficient than class B, but still much more efficient than class A. Recall that the maximum theoretical efficiency for class A is 25 per cent. Class B is 78 per cent, and for class AB 70 per cent would be a likely figure, but it of course depends on just how much forward bias is used.

7-8 Practical Push-Pull Amplifiers

All the PA circuits thus far discussed are in general use, as well as many other variations of these circuits. However, they all have certain disadvantages compared to the complementary amplifier shown in Fig. 7-19. Conse-

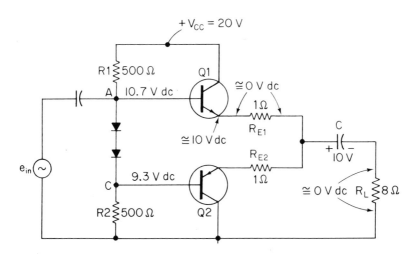

FIG. 7-19. Complementary symmetry class AB push-pull amplifier

quently, it is this basic configuration that is being used as the most economical means of providing a given amount of power output in modern designs. The disadvantages that it overcomes include

1. Only one power supply is necessary.
2. No requirement of transformer coupling.
3. No paraphase amplifier requirement.

Proper operation of this circuit requires that point B be at a dc potential of about one half the supply voltage, V_{CC}. (Refer to Fig. 7-19.) In this case that means 10 V at point B and hence 10.7 V at point A, assuming silicon transistors and a class AB bias on $Q1$ of $V_{BE} = 0.7$ V. The two silicon diodes are always forward biased and drop a relatively constant amount of voltage, equal to about 1.4 V. They are thus providing the necessary small amount of forward bias for $Q1$ and $Q2$ to allow class AB operation and hence eliminate crossover distortion. If the diodes are physically located in the same thermal environment as $Q1$ and $Q2$, temperature effects on the idle (or quiescent) current will be minimized. The *idle current* is the dc current flowing in the collectors of the transistors with no ac input signal. For example, a temperature rise would cause the forward voltage drop V_{BE} of $Q1$ and $Q2$ to decrease and therefore increase the dc collector current of each transistor. However, the diodes' forward voltage drop would also decrease, and tend to decrease I_B and hence I_C. Thus an equilibrium condition exists. In some designs a single diode and series resistor are used, or even just a resistor in place of the two series diodes.

The emitter resistors R_{E1} and R_{E2} have very low resistances and serve to stabilize the transistors from thermal runaway and also to ensure that one transistor is off while the other is on (except during crossover). They typically are 1 Ω resistors or even a fuse that can also provide overload protection. Because of their low value and low dc emitter currents (typically 10 mA of idle current), the dc voltage drop across them may be ignored.

The capacitor C serves to keep direct current out of the load, and even more importantly, since it is charged to 10 V with the polarity as shown in Fig. 7-19, it serves as the dc supply during the conduction of $Q2$. Since this is normally a high current draw, it must be a large electrolytic capacitor to enable storage of enough energy.

Resistors $R1$ and $R2$ (500 Ω in this case) should be equal to one another to allow the bias voltages indicated in Fig. 7-19. They must be small enough to provide enough forward bias current for $Q1$ and $Q2$ to minimize crossover distortion. If the necessary dc base current for sufficient forward bias is 0.1 mA for each transistor (a typical value), then the current through $R1$ should at least be 10 times that amount to keep from loading down the base biases. Of course, using too low a value for $R1$ and $R2$ causes an excessive power drain from the dc supply and results in an overall decrease in efficiency, thus lowering the circuit input impedance.

We shall now analyze what happens when a 10 V p-p signal is applied to the circuit of Fig. 7-19. Initially consider the point in time when the input is at its maximum +5 V peak value. Point A, which had been at 10.7 V, will now have an instantaneous value of 15.7 V, and point C will be at 14.3 V, an increase of 5 V from its quiescent value of 9.3 V. $Q1$ becomes strongly forward biased, and its emitter voltage will jump from its approximate 10 V level to 15 V. With the capacitor C appearing as a short circuit to this ac voltage change, the sudden 5 V jump at the emitter of $Q1$ results (by voltage divider action between R_{E1} and R_L) in an output voltage change from zero to 5 V × (8 Ω/1 Ω + 8 Ω) ≃ 4.5 V. Point B has jumped from 10 to 14.5 V(15 V + $V_{R_{E1}}$) and the base of $Q2$ is at 14.3 V, so $Q2$ is nonconducting, as desired, while $Q1$ is conducting. Figure 7-20 summarizes these voltage changes for the +5 V instantaneous input signal with the new instantaneous voltage levels shown enclosed.

Applying similar logic to the reverse half-cycle of the input voltage results in a −4.5 V level across R_L when $e_{in} = -5$ V with $Q2$ now conducting and $Q1$ turned off. Remember that the current drawn through the load (4.5 V

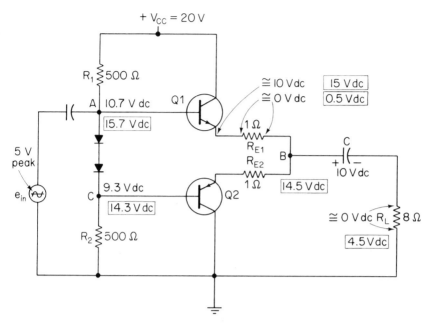

FIG. 7-20. Amplifier response at instant $e_{in} = +5V$

/8 Ω $\simeq$ 0.55 A when $e_{in} = -5$ V, its peak value) cannot be supplied by the 20 V supply since $Q1$ is "turned off." As previously mentioned, the charge on capacitor C is used to supply this energy. The lower the frequency of the input signal, the longer it must be able to supply energy and the greater its capacitance must be. We shall show how to determine an adequate value in Section 8-3, but typically 500–1000 μF is required to allow an output to a frequency down to 20 Hz.

The maximum peak-to-peak ac input voltage for this amplifier is approximately 10 per cent less than the supply voltage, or 18 V p-p in this case. Clipping will occur otherwise; therefore, the maximum output power would be (18 V p-p/$2\sqrt{2}$)^2/R_L or about 5 W *rms* for the 8 Ω load. The voltage drop across the emitter resistors decreases this somewhat. To obtain greater output power would require a higher supply voltage. The power requirement of $Q1$ and $Q2$ at maximum output is about equal to

$$P_{Q1\ max} = P_{Q2\ max} \simeq \frac{V_{CC}^2}{8\pi^2 R_L}$$

and in this case would be

$$P_{Q1\ max} = \frac{(20\ \text{V})^2}{4\pi^2(8\ \Omega)} = 1.26\ \text{W}$$

Thus relatively low power transistors are able to deliver significant power to the load and heat sinking is kept to a minimal level. At the no-input-signal condition the transistor power dissipation is about equal to $V_{CC}/2$ times the idle current, in this case (10 V)(10 mA) = 100 mW. The power required from the power supply at maximum output is about

$$P_{\text{supply max}} \simeq \frac{V_{CC}/2}{2\pi R_L} \qquad (7\text{-}15)$$

In practice a driver stage is required for the amplifier of Fig. 7-20 since it requires a high-voltage ac signal at a significant current level because of its low input impedance. (We shall discuss methods of driving this amplifier in Chapter 9). The input impedance seen by e_{in} of Fig. 7-19 is approximately

$$R1 \parallel R2 \parallel h'_{fe}(R_L + R_E)$$

and thus for our case, with $h'_{fe_1} = h'_{fe_2} = 100$,

$$R_{\text{IN}} \simeq 500 \parallel 500 \parallel 100(1 + 8)$$
$$\simeq 196 \, \Omega$$

It is normal to drive a complementary symmetry amplifier with an emitter follower to provide an input impedance high enough not to load down a previous voltage amplifier.

PROBLEMS

1. What is the difference between large-signal amplifiers and power amplifiers?
2. A 2N3766 transistor is to be used at a 100°C maximum case temperature. Determine the allowable power dissipation.
3. Determine the heat-sink size requirement to allow 5 W dissipation for a 2N3766 transistor. Assume that $\theta_{CS} = 0.4°C/W$, $T_{J\,\text{max}} = 175°C$, $\theta_{JC} = 7.5°C/W$, and the ambient temperature is 90°C.
4. If the transistor–heat sink junction of Example 7-2 were changed to direct (no mica washer) with silicone grease, determine the reduced size requirement for the heat sink.
5. What factors make heat-sink design an art instead of a science?
6. An amplifier has a transfer curve as shown in Fig. 7-5 except all values for I_C are doubled. Determine the percentage of the second, third, and fourth harmonics, and total distortion.
7. Ten per cent negative feedback ($B = -0.1$) is introduced to the amplifier

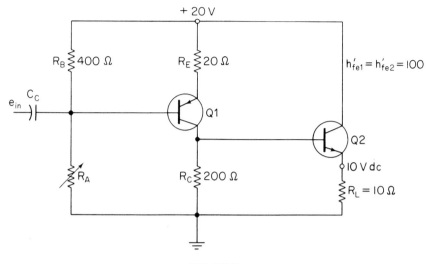

FIG. P7-8.

in Problem 6. Determine the new value of total distortion ($R_L = 5$ kΩ).

8. In the circuit shown in Fig. P7-8, determine G_{V1}, G_{V2}, G_{voa}, Z_{in}, G_{ioa}, and G_{poa}.
9. In the circuit of Fig. P7-8, determine the maximum possible value of e_{in} that will not cause clipping of the output signal. With that level of input voltage, determine the ac load power and η.
10. Calculate the approximate required value for R_A in the circuit of Fig. P7-8.
11. What are the disadvantages of CE circuits compared to CC for class A power amplifiers?
12. The circuit of Fig. P7-8 is to drive a transformer-coupled 300 Ω load to a 150 V p-p level. Determine the required transformer turns ratio and reflected impedance when the ac emitter voltage is 20 V p-p.
13. Determine G_{voa} for Problem 12.
14. What are the basic limitations of transformer coupling in PA applications?
15. Give two reasons for the good efficiency of a class B amplifier.
16. An amplifier is to provide 200 W of ac output power. Give the approximate transistor power rating required for a class A and class B design.
17. Discuss the cause of crossover distortion and how it is prevented.
18. Determine the instantaneous voltages that result in Fig. 7-20 when the input signal is a -5 V instantaneous value.
19. Determine the supply voltage required for the complementary symmetry amplifier of Fig. 7-20 to deliver 25 W to the 8 Ω load. What values would then be most appropriate for $R1$ and $R2$?
20. Calculate the required power rating of $Q1$ and $Q2$ for Problem 19. Determine the required wattage rating for the power supply.

8

Amplifier Frequency Response

8-1 Introduction

Our study of amplifiers up to this point has mainly concerned itself with only one range of frequencies. That is the midfrequency range, where changes in ac input frequency had little or no effect on our analysis. Figure 8-1 shows a

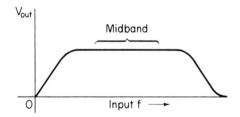

FIG. 8-1. Typical amplifier frequency-response curve

typical frequency-response curve for an amplifier. The input signal is assumed to be of constant amplitude and is varied from 0 Hz (dc) up to a very high frequency where the amplifier no longer has any appreciable output. The result is an area of relatively flat response, known as the *midband*, and two areas of reduced response: one at low frequencies and one at high frequencies.

It is these two regions of amplifier performance that form the basis of our study in this chapter.

The low-frequency response is determined by emitter or source bypass capacitors and the coupling method between the signal source, amplifier stages, and load. If direct coupling is used throughout, and no bypass capacitors are used, the amplifier should respond down to 0 Hz and is then called a dc *amplifier*. The other possible forms of coupling cause a reduction in low-frequency gain in a predictable fashion. The most often used methods of coupling are capacitive coupling and transformer coupling. Transformer coupling also limits the high-frequency gain.

Since transformers are now seldom used for coupling, the major cause of high-frequency loss of gain is the inherent junction capacitances of the active amplification devices—bipolar junction transistors (BJTs) or FETs. These small capacitances tend to develop a low enough reactance at high frequencies to effectively short out a portion of the signal, thereby reducing the gain. Another source of high-frequency attenuation is the stray wiring capacities inherent in any circuit, and they have the same effect as the junction capacitances.

Amplifiers that have midbands in the approximate region of 20 Hz to 20 kHz are termed *audio amplifiers*. They are used to amplify signals in the range of frequencies heard by the human ear, such as speech and music. A *video amplifier* must have a very wide frequency response—from very low frequencies (often down to direct current) on up to several hundred kilohertz or more. A *radio-frequency amplifier* (RF amplifier) is operated at only high frequencies. Typical frequencies range from 100 kHz up to 10,000 MHz. Those amplifiers that are operated in only a very small frequency range in the RF spectrum are termed *tuned amplifiers*. They have special properties that reject all frequencies except a narrow desired range. This is usually accomplished through the use of LC tuned circuits or active-device solid-state filters.

8-2 Capacitive Filters

Since virtually all the high and low frequency effects in today's amplifiers are caused by capacitance, a review of the basics involved is in order. First consider the circuit shown in Fig. 8.2a. At very high input frequencies the capacitor C appears as a short circuit, since $X_C = 1/2\pi fC$: therefore, increases in f cause a decrease in X_C. As the frequency goes down, however, X_C becomes larger, and eventually the voltage divider action of C and R begins to attenuate the output voltage (see Fig. 8-2b). Of particular interest to us is the output

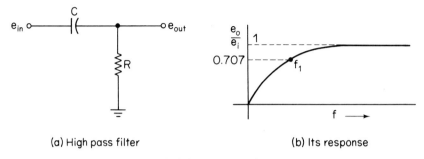

(a) High pass filter (b) Its response

FIG. 8-2. High pass filter

voltage when $X_C = R$. By the voltage divider rule

$$e_o = e_i \times \frac{R}{R - jX_C}$$

If $R = X_C$, then

$$e_o = e_i \frac{R}{R - jR} = e_i \frac{R}{\sqrt{2}\,R}$$
$$= 0.707 e_i$$

We see that when $R = X_C$ the output voltage is 0.707 of the input voltage, and by definition we shall call this the *low-frequency cutoff*, f_1. Figure 8-2b shows a graph of e_o/e_i versus frequency for a high pass filter. Since by definition f_1 is the frequency where $X_C = R$, we can say that

$$R = \frac{1}{2\pi f_1 C}$$

or

$$f_1 = \frac{1}{2\pi RC} \tag{8-1}$$

Thus the low-frequency cutoff of a high pass filter can be easily calculated by Eq. (8-1).

In many cases the response of an *RC* circuit or total amplifier is expressed in terms of decibels (dB) instead of just an ordinary gain or attenuation ratio. In terms of the ratio of two powers, a decibel is defined as

$$\text{decibel} = 10 \log_{10}\left(\frac{P_2}{P_1}\right) \tag{8-2}$$

The logarithm in Eq. (8-2) is to the base 10. Since power is related to the square of voltage and current, it follows that

$$\text{decibel} = 20 \log_{10}\left(\frac{V_2}{V_1}\right) \quad (8\text{-}3)$$

and

$$\text{decibel} = 20 \log_{10}\left(\frac{I_2}{I_1}\right) \quad (8\text{-}4)$$

since the log $V^2 = 2 \log V$. The following relationship is also useful:

$$\log \frac{1}{X} = -\log X \quad (8\text{-}5)$$

We could therefore calculate decibels down at the low-frequency cutoff f_1 as

$$\text{decibel} = 20 \log \frac{1}{0.707} = 20 \log 1.414$$
$$\cong 3 \text{ dB reduction} \quad \text{or} \quad -3 \text{ dB}$$

Thus our low-frequency cutoff frequency is also the 3 dB "down" frequency, as well as being the point where power to the load is one half of its maximum value. Since $P = V^2/R$, then $(0.707\text{ V})^2$ is half the original power. Thus at the low-frequency cutoff, f_1, the power is reduced to one half its original value, and the voltage is reduced to 0.707 of its original value.

EXAMPLE 8-1

Determine f_1 for the circuit shown in Fig. 8-3. How many decibels down will the output be at one tenth of f_1?

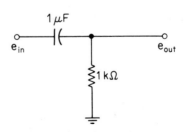

FIG. 8-3. Example 8-1

Solution:

$$f_1 = \frac{1}{2\pi RC} = \frac{0.159}{10^3 \times 10^{-6}} = 0.159 \times 10^3$$
$$= 159 \text{ Hz}$$

At f_1 or 159 Hz the value of X_C is equal to R, or 1 kΩ. Therefore, at $\frac{1}{10}f_1$, X_C will be 10R or 10 kΩ. Therefore, the output voltage at $f_1/10$ is

$$e_{out} = e_{in}\frac{R}{R - jX_C}$$

The ratio of e_o/e_{in} is

$$\frac{e_{out}}{e_{in}} = \frac{1 \text{ k}\Omega}{1 \text{ k}\Omega - j10 \text{ k}\Omega} \cong 0.1$$

and thus, from Eq. (8-5),

$$\text{decibel} = 20 \log 0.1 = -20 \log \frac{1}{0.1}$$
$$= -20 \log 10 = -20 \text{ dB}$$

The low pass filter shown in Fig. 8-4a has the characteristics shown in part b of the figure. It has a high-frequency cutoff, f_2, as shown at 0.707 of its

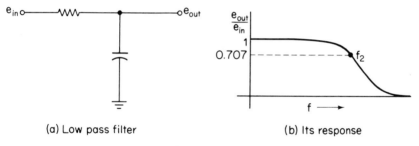

(a) Low pass filter (b) Its response

FIG. 8-4. Low pass filter

maximum. Since the capacitor's reactance keeps decreasing as the frequency goes up, the output will drop by voltage divider action. The high cutoff frequency, f_2, also termed the 3 dB down frequency, half-power point, or break frequency, will occur when $X_C = R$ just as for the high pass filter.

EXAMPLE 8-2

Determine f_2 for the circuit of Fig. 8-4 if $R = 1$ kΩ and $C = 0.01$ μF.
Solution:

$$f_2 = \frac{1}{2\pi RC} = \frac{0.159}{10^3 \times 10^{-8}}$$
$$= 0.159 \times 10^5$$
$$= 15.9 \text{ kHz}$$

If the circuits of Examples 8-1 and 8-2 were cascaded together, a band of frequencies would be passed, with the very low and high frequencies attenuated. The band of frequencies passed between f_1 and f_2 is the *bandwidth* of the circuit, which is appropriately called a bandpass filter. The circuit and its response are shown in Fig. 8-5.

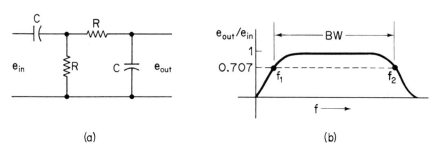

FIG. 8-5. Bandpass filter

8-3 Amplifier Performance at Low Frequencies

As previously mentioned, the low-frequency response of an amplifier is usually caused by the coupling capacitor between stages or between the input and/or output and the amplifier. Another source of loss at low frequencies is the emitter or source bypass capacitor, which develops appreciable impedance at low frequencies and thus no longer is a complete bypass. Since the amplifying device itself is not responsible for this loss, this discussion will be applicable to amplifiers with any type of active device.

In general, the easiest means of determining f_1 due to a coupling capacitor is to find the frequency at which X_C is equal to the resistance seen from the capacitor's two terminals. That frequency is $1/2\pi RC$, as previously shown. Consider the circuit of Fig. 8-6. It is seen that three different capacitors, C_{C1}, C_{C2}, and C_E can affect the ultimate value of f_1 for this amplifier. We shall analyze each one separately and then determine the overall value of f_1.

The 1 μF coupling capacitor, C_{C1}, "sees" the following resistances from its terminals: R_S in series with the parallel combination of R_A, R_B, and R_{in}. Remember that the voltage sources should be considered as short circuits to alternating current. Figure 8-7 should help in your visualization of the resistance seen from the C_{C1} terminals. The resistance seen looking into $Q1$'s base, R_{in}, is equal to approximately $h'_{fe} h_{ib}$ if C_E is considered as a short circuit, and hence the resistance R seen by C_{C1} is

$$R = R_S + R_A \| R_B \| h'_{fe} h_{ib}$$
$$\simeq 3 \text{ k}\Omega + 10 \text{ k}\Omega \| 100 \text{ k}\Omega \| 2.6 \text{ k}\Omega$$

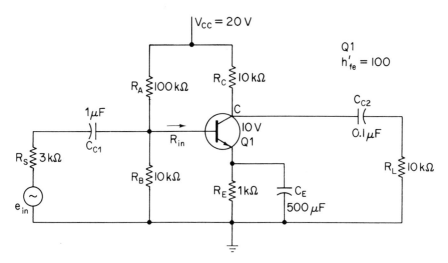

FIG. 8-6. Amplifier for low-frequency analysis

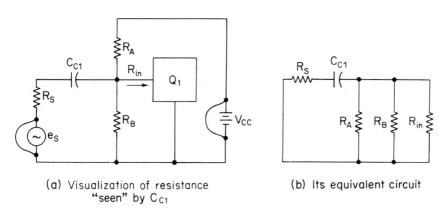

(a) Visualization of resistance "seen" by C_{C1}

(b) Its equivalent circuit

FIG. 8-7.

since

$$h_{ib} = \frac{0.026}{I_E} = \frac{0.026}{1 \text{ mA}} = 26 \, \Omega.$$

Therefore

$$R \simeq 3 \text{ k}\Omega + 2 \text{ k}\Omega = 5 \text{ k}\Omega$$

and the f_1 for capacitor C_{C_1} is

$$f_{1C_{C1}} = \frac{1}{2\pi C_{C_1} R} = \frac{1}{2\pi \times 10^{-6} \times 5 \times 10^3}$$
$$= 32 \text{ Hz}$$

The resistance "seen" by C_{C_2} in Fig. 8-6 is R_L in series with R_C in parallel with the resistance of the reverse-biased collector junction. The reverse-biased junction impedance is typically 100 kΩ or more and is therefore safe to ignore. Thus $R_L + R_C$ in series $= 10\text{ k}\Omega + 10\text{ k}\Omega = 20\text{ k}\Omega$. The f_1 for C_{C_2} is

$$f_{1C_{C_2}} = \frac{1}{2\pi RC} = \frac{1}{2\pi(20 \times 10^3) \times 0.1 \times 10^{-6}}$$
$$= 32\text{ Hz}$$

The value of f_1 due to C_{CE} can also be determined by finding the value of resistance it "sees" across its terminals. It "sees" R_E in parallel with the combination of h_{ib} plus the impedance looking from inside the transistor out the base lead divided by h'_{fe}. This seems reasonable since looking into the base lead all the emitter impedances are multiplied by h'_{fe}. Figure 8-8

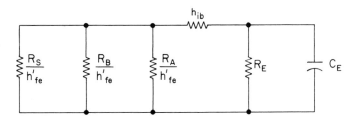

FIG. 8-8. Resistance "seen" by C_E

illustrates these effects. Thus C_E "sees"

$$R = R_E \left\| \left(26\text{ }\Omega + \frac{R_A \| R_B \| R_S}{h'_{fe}}\right)\right.$$

Hence

$$R \simeq 1\text{ k}\Omega \left\| \left(26\text{ }\Omega + \frac{10\text{ k}\Omega \| 100\text{ k}\Omega \| 3\text{ k}\Omega}{100}\right)\right.$$
$$= 1\text{ k}\Omega \| (26\text{ }\Omega + 22.5\text{ }\Omega)$$
$$= 48\text{ }\Omega$$

Therefore,

$$f_{1 C_E} = \frac{1}{2\pi(48\text{ }\Omega)(500 \times 10^{-6})}$$
$$\simeq 6\text{ Hz}$$

Since 6 Hz is so much smaller than 32 Hz, it is safe to assume that the two 32 Hz frequencies will determine f_1 with C_E having little effect. Whenever two or more effects are causing about the same value of f_1 in an amplifier system, the following relationship may be used to determine the overall response:

$$f_1 \text{ cutoff overall} \cong f_1(1.1)^n \qquad (8\text{-}6)$$

where f_1 is the cutoff frequency of n different effects. In our case f_1 is 32 Hz and n is 2. Therefore the overall cutoff frequency for the amplifier of Fig. 8-6 is

$$32 \text{ Hz}(1.1)^2 = 38 \text{ Hz}$$

If the value of f_1 due to C_{C1} and C_{C2} had been 30 and 34 Hz, respectively, the overall value of f_1 would be about the same, since the average of these two is still 32 Hz, and the average value of cutoffs can be used in Eq. (8-6) if they are quite close to one another (within 10-20 per cent).

If the value of C_E in this amplifier had been 50 μF instead of 500 μF, $f_{1\,CE}$ would have been 60 Hz instead of 6 Hz. Thus the overall low-frequency response of this amplifier would have been determined by C_E, it causing a much higher f_1 than the coupling capacitors. A good low-frequency response requires a large value of bypass capacitance. This factor as well as the input impedance variance with h_{fe} has resulted in minimal use of bypass capacitors in modern designs.

EXAMPLE 8-3

Determine f_1 for the amplifier shown in Fig. 8-9.

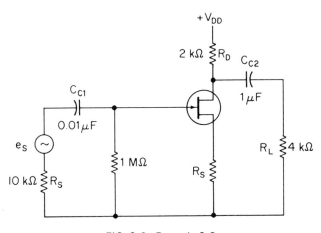

FIG. 8-9. Example 8-3

Solution:
The input coupling capacitor C_{C1} sees 10 kΩ + (1 MΩ in parallel with R_{in} of the FET). Since R_{in} is very large, the resistance seen by the capacitor is just

$R = (10 \text{ k}\Omega + 1 \text{ M}\Omega) \cong 1 \text{ M}\Omega$, and then f_1 for C_{C1} is

$$f_1 = \frac{1}{2\pi CR}$$

$$= \frac{1}{2\pi(0.01 \times 10^{-6})(10 \text{ k}\Omega + 1 \text{ M}\Omega)}$$

$$= \frac{10^8}{2\pi \times 10^6}$$

$$= \frac{100}{2\pi} \simeq 16 \text{ Hz} \quad \text{due to } C_{C1}$$

The output coupling capacitor C_{C2} "sees" $2 \text{ k}\Omega + 4 \text{ k}\Omega$ in series, and its cutoff frequency is

$$f_1 = \frac{1}{2\pi \times 10^{-6} \times 6 \times 10^3}$$

$$= \frac{0.159 \times 10^6}{6 \times 10^3}$$

$$= 26.5 \text{ Hz} \quad \text{due to } C_{C2}$$

Since this is almost double the 16-Hz cutoff, it is safe to say that the overall low frequency cutoff is around 26 Hz.

8-4 High-Frequency Response of the FET

The high-frequency response of an FET amplifier is determined by the internal capacitances between the FET's three terminals. They are C_{gs}, which is from gate to source, and C_{gd} from gate to drain. They have typical values of from 1–10 picofarads (pF). The capacitance from drain to source, C_{ds} is small, typically 0.1 pF, and is usually ignored. C_{gd} is often listed in manufacturers' data sheets as C_{rss}, the *reverse transfer capacitance*. The sum of C_{gd} and C_{gs} is often listed as C_{iss}, the small-signal, common-source, short-circuit input capacitance:

$$C_{iss} = C_{gd} + C_{gs} \tag{8-7}$$

At first glance then, the high-frequency cutoff calculation for the amplifier of Fig. 8-10 appears quite straightforward. Assuming values of 5 pF for C_{gd} and C_{gs}, the input shunt capacitance can be reasoned to be C_{gd} in parallel with C_{gs}, or 10 pF. This is reasoned by assuming R_D is a relatively low impedance and effectively shorts C_{gd} to ground as shown in Fig. 8-10. The resistance seen by this shunt capacitance is $R_S \| R_G$ or 0.5 MΩ. Hence f_2 is

$$f_2 = \frac{1}{2\pi(0.5 \times 10^{+6})(10 \times 10^{-12})}$$

$$\simeq 38 \text{ kHz}$$

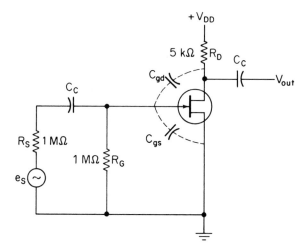

FIG. 8-10. High-frequency effects of FET

Unfortunately, this result is incorrect due to the Miller effect. The *Miller effect* is an increase of input capacitance due to multiplication of C_{gd} by a factor of 1 plus the voltage gain of the stage, or $C_{in} = C_{gd}(1 + A_v)$.

With reference to Fig. 8-11, assume a voltage gain of A_v and recall the phase inversion from input to output of such an amplifier. Since the input

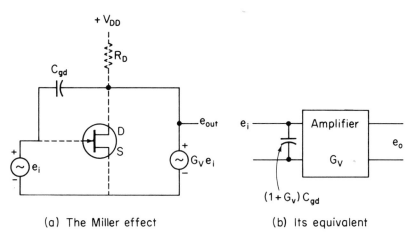

(a) The Miller effect   (b) Its equivalent

FIG. 8-11.

and output voltages are then series aiding, the current through C_{gd} will be $(1 + A_v)$ times larger than if the input generator were driving C_{gd} alone. This current draw factor of $(1 + A_v)$ essentially means that C_{gd} looks like a capacitor whose value is $(1 + A_v)C_{gd}$, and this is the Miller effect. This capacitor appears from the gate to ground. This is the same effect we discussed earlier (Section 5-8) when a feedback resistor was connected from the output to the input of a phase-inverting amplifier. Once again the input current is increased by the $(1 + A_v)$ factor, which means an effective reduction in the value of resistance by the $(1 + A_v)$ factor. Note that an increase in capacitance or decrease in resistance corresponds to a greater current draw from an ac voltage source.

The Miller effect is a very important factor in FET amplifier high-frequency response. It also affects f_2 for all other phase-inverting amplifiers, such as BJT CE amplifiers covered in Section 8-6.

EXAMPLE 8-4

Determine G_v and f_2 for the amplifier shown in Fig. 8-12.

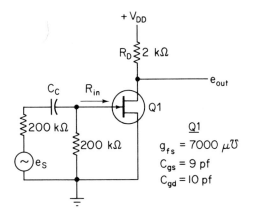

FIG. 8-12. Example 8-4

Solution:

$$A_v \simeq g_{fs}R_D$$
$$= 7000 \, \mu\mho \times 2 \, k\Omega$$
$$= 14$$

Therefore, the input capacitance is

$$C_{in} = C_{gs} + (1 + A_v)(C_{gd})$$
$$= 9 \, pF + 15(10pF) = 159 \, pF$$

The resistance seen by this shunt input capacitance is $200 \, k\Omega \, || \, 200 \, k\Omega \, || \, R_{in}$, which

≈ 100 kΩ since R_{in} is very large. Therefore, f_2 is

$$f_2 = \frac{1}{2RC} = \frac{0.159}{159 \times 10^{-12}(100 \text{ k}\Omega)}$$
$$= \frac{0.159 \times 10^7}{159} = \frac{159 \times 10^4}{159}$$
$$= 10 \text{ kHz}$$

The previous example, although slightly exaggerated, illustrates the relatively poor frequency response of high-gain FETs, especially when driven from high impedance sources.

Also note that better high-frequency response results when low values of R_G are used. Hence when an extremely large input impedance is not required of an FET stage, a low value of R_G will serve to extend the high-frequency response.

The high-frequency response of the source-follower configuration is superior to the common-source variety, since the only capacitance shunting out the input signal is C_{gd}. The effective value of C_{gs} is reduced to almost nothing by the 100 per cent feedback effects between its two terminals.

8-5 High-Frequency Response of the BJT

The high-frequency response of a bipolar junction transistor (BJT) amplifier stage is determined by a number of complex factors. More often than not, the designer will breadboard the design in the laboratory to determine these effects experimentally for a given design and validate his calculations. The high-frequency cutoff is mainly due to the inherent internal junction capacitances, which are in turn complexly related to transistor manufacturing techniques, the way in which the transistor is operated (bias conditions) and, to a lesser extent, changes in temperature. Another important effect is the amount of voltage gain provided by the amplifier, since the Miller effect, explained in Section 8-4, is also a definite factor here. One last consideration is the shunt stray wiring capacitance, which exists to some degree in every circuit.

Figure 8-13 shows a CE amplifier with the significant shunt capacitances shown in dashed lines. The inherent collector–base junction capacitance $C_{b'c}$ is often given as C_c, C_o, C_{CB}, or C_{OB} on manufacturers data sheets. Its value is dependent on manufacturing techniques and is also inversely proportional to the square root of the reverse-biased collector–base voltage. Typical values for it range from 1 pF for transistors especially designed for high-frequency operation up to about 50 pF for some audio power transistors. The capaci-

184 *Amplifier Frequency Response*

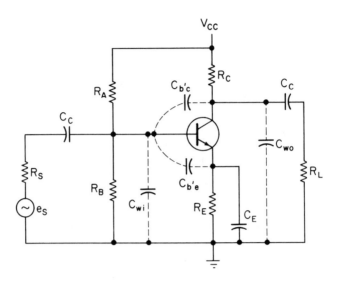

FIG. 8-13. Common-emitter amplifier; $C_{b'c}$, $C_{b'e}$, C_{wi}, and C_{wo} affect high-frequency response

tance $C_{b'e}$ is composed of the emitter-to-base junction capacitance and the emitter-to-base diffusion capacitance. The junction capacitance is quite constant, whereas the diffusion capacitance depends upon the number of charges in and near the depletion region and is directly proportional to the dc emitter current, I_E. Values of from 10 pF up to 5000 pF are typical for this capacitance. The capacitances C_{wi} and C_{wo} are the stray wiring capacitances on the input and output of the amplifier. Values of 10 pF each are typical, but they vary widely depending on circuit layout.

It is seen that the shunting capacitances of a CE amplifier are many and difficult to predict. Fortunately, a large number of linear amplifiers are of the audio variety in which these effects can often be ignored. An exception to this occurs when an extremely high voltage gain stage is utilized with a transistor having large junction capacitances. The effective multiplication of $C_{b'c}$ by $(1 + A_v)$ due to the Miller effect may cause a 10 kHz cutoff, or even lower, which is not acceptable for many amplifiers. In those amplifiers in which high-frequency operation is required, the approximate analysis presented in the following paragraphs should be accompanied by laboratory testing. Only by experimental test can all the complex high-frequency effects be fully accounted for.

Much developmental work has been accomplished in recent years on high-frequency transistors. They are now available for operation as high as 10 gigahertz (GHz) (1 GHz = 1000 MHz) at power levels up to 10 and 20 W. This level of performance is of value in radio communications equipment.

Naturally, the cost of such a device is many times greater than an audio transistor at 10 or 20 W with the cost rising exponentially with increases in required operating frequency. These cost considerations make selection of the proper transistor for a given circuit important. As an aid to the designer, manufacturers' data sheets provide information that makes approximate calculations of high-frequency cutoff possible. Older specifications showed the alpha and beta cutoff frequencies, respectively, f_α and f_β; the first is the frequency at which α (h_{fb}) falls by 3 dB and the second is the frequency at which β (h_{fe}) falls by 3 dB due to the junction capacitances on the input of the transistor, $C_{b'e}$, and $(1 + A_v)C_{b'c}$. It can be shown that f_α and f_β are related as

$$f_\alpha = \beta f_\beta \qquad (8\text{-}8)$$

It follows from this that the CB amplifier has a generally higher f_2 than does a CE amplifier.

These frequencies are no longer in common use, having been replaced by the *current gain bandwidth frequency*, f_T. As an approximation,

$$f_T \simeq \frac{f_\alpha}{1.2} \qquad (8\text{-}9)$$

Up to a point, f_T is proportional to both dc collector voltage and collector current and reaches a maximum for typical high-frequency transistors at $V_{CE} = 10$ V and I_C in excess of 20 mA. Given f_T, the designer can calculate the approximate value of f_2 due to the input capacitances of a zero-feedback CE amplifier as f_β, where

$$f_2 \simeq f_\beta \simeq \frac{f_T}{1.2 h_{fe}} \qquad (8\text{-}10)$$

Since f_T is the current gain bandwidth product, it is the frequency at which the amplifier current gain (h_{fe}) has fallen to 1. Hence it is a much higher frequency than f_2 (f_2 is the frequency at which G_v has fallen to 0.707 of its midband value). Operation with an h_{fe} of 1 is common, however, in some high-frequency circuits in which power gains greater than 1 still result (since G_v is still > 1); they are therefore useful circuits. To determine f_2 in an amplifier incorporating negative feedback, the following formula should be utilized:

$$f_{2fb} = f_2(1 - BG_v) \qquad (8\text{-}11)$$

The term f_{2fb} is the high-frequency cutoff with negative feedback, B is the feedback factor, G_v is the voltage gain without feedback, and f_2 is the high-frequency cutoff without feedback. Equation (8-11) shows that extended high-frequency response results from the use of negative feedback. Feedback also serves to improve the low-frequency response of an amplifier, as predicted by

$$f_{1fb} = \frac{f_1}{1 - BG_v} \qquad (8\text{-}12)$$

If the driving source impedance is low, the value of f_2 predicted as previously outlined becomes meaningless, since the generator no longer supplies a constant current to the changing (with frequency) input impedance of the amplifier. It is then necessary to analyze the circuit with respect to the values of $C_{b'c}$ and $C_{b'e}$, which are normally provided by the manufacturer.

For example, the circuit of Fig. 8-14 has $C_{b'c} = 10$ pF and $C_{b'e} = 1000$

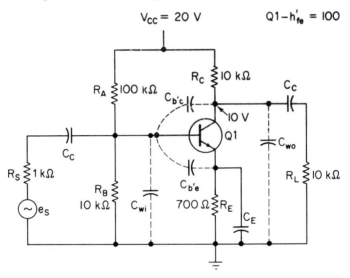

FIG. 8-14. Common-emitter amplifier

pF. If we assume an input wiring capacitance of 20 pF, the total capacitance shunting the *input* signal to the amplifier is

$$C_{\text{shunt total}} = C_{wi} + C_{b'e} + (1 + G_v)C_{b'c}$$

The voltage gain $G_v \simeq R_C \| R_L / h_{ib} = 5 \text{ k}\Omega / 26 \ \Omega \simeq 200$. Therefore, the total shunt input capacitance is approximately

$$20 \text{ pF} + 1000 \text{ pF} + (200)10 \text{ pF} \simeq 3000 \text{ pF}$$

The shunt resistance seen by this shunt capacitance is

$$R_A \| R_B \| R_S \| h'_{fe} h_{ib}$$

and equals
$$R_{shunt} = 10 \text{ k}\Omega \,||\, 100 \text{ k}\Omega \,||\, 1 \text{ k}\Omega \,||\, 2.6 \text{ k}\Omega \cong 670 \,\Omega$$

Therefore, f_2 is
$$f_2 = \frac{1}{2\pi C_{shunt} R_{shunt}}$$
$$= \frac{1}{2\pi \times 3000 \text{ pF} \times 670 \,\Omega}$$
$$\simeq 80 \text{ kHz}$$

If this amplifier were operated without the emitter bypass capacitor, negative feedback would reduce the G_v to $G'_v \simeq R_L/R_E = 10 \text{ k}\Omega/700 \simeq 14.3$. Then $C_{shunt\,total}$ would equal

$$20 \text{ pF} + 1000 \text{ pF} + (14.3)10 \text{ pF} \simeq 1160 \text{ pF}$$

and f_2 would rise to about 200 kHz.

Because of the severe effect on f_2 due to Miller effect multiplication of $C_{b'c}$, RF amplifiers do not normally operate in the emitter-bypassed high-voltage-gain configuration.

One last effect on the actual value of f_2 is due to the output circuit, which effectively has $C_{b'c}$ and C_{wo} shunted with $R_L \,||\, R_C$ to form a low pass filter, and thus f_2 due to the output effects is

$$f_2 = \frac{1}{2\pi(C_{b'c} + C_{wo})(R_L \,||\, R_C)} \tag{8-13}$$

which in the case of Fig. 8-14 provides an f_2 with $C_{wo} = 10 \text{ pF}$ of

$$f_2 = \frac{1}{2\pi(20 \text{ pF})5 \text{ k}\Omega}$$
$$= 1.5 \text{ MHz}$$

This means that the high-frequency cutoff f_2 is determined solely by the input effects. Thus f_2 for the amplifier of Fig. 8-14 is 80 kHz bypassed or 200 kHz unbypassed.

If f_2 for the input and output shunt capacitance effects are roughly equal, the overall value of f_2 is predicted by

$$f_{2\,cutoff\,overall} = \frac{f_2}{(1.1)^n} \tag{8-14}$$

where f_2 is the average high-frequency cutoff of n different effects. The individual values of f_2 should be within 10–20 per cent of one another before the

188 Amplifier Frequency Response

use of Equation (8-14) is necessary. Of course, if one effect is very much lower than the others, it is the dominant factor. This equation applies to the input and output effects of a single stage, as well as any number of transistor stages cascaded together.

8-6 Review of BJT High-Frequency Effects

In review, the following steps should be taken to determine the value of f_2 for a single-stage BJT amplifier:

1. If the driving source impedance is high (at least five times the overall amplifier input impedance) and f_T is given, f_2 can be approximated as f_T/h_{fe} for a non-feedback amplifier or $(f_T/h_{fe})/(1 - BG_v)$ with feedback, where B is the feedback factor and G_v the open-loop gain.

or

2. If the driving source impedance is low (or if a slightly more accurate value of f_2 is desired than the one shown by step 1), calculate f_2 as equal to $1/(2\pi C_{shunt} R_{shunt})$. C_{shunt} equals the sum of the input wiring capacitance plus $C_{b'e}$ plus $(1 + G'_v)C_{b'c}$, where G'_v is the actual voltage gain of the transistor from input (base) to output (collector). R_{shunt} is the total effective resistance seen from the two shunt capacitance terminals.

and then

3. Calculate f_2 due to the output capacitance effects using the same formula as in step 2, where C_{shunt} now equals $C_{b'c} + C_{wo}$, and R_{shunt} usually equals $R_C \| R_L$.

and then

4. The total composite high-frequency cutoff due to both input and output effects should be determined using Eq. (8-14) if they are close in value. If they are of widely different values, f_2 is equal to the lower one.

EXAMPLE 8-5

The amplifier shown in Fig. 8-15 is to be analyzed for its high-frequency cutoff, f_2.

Solution:
Using the steps just outlined,
1. R_S is much greater than the overall input resistance, and therefore f_2 due to

$h_{fe} = 50$

$C_{wi} = 20$ pF

$C_{wo} = 245$ pF

$C_{b'e} = 500$ pF

$C_{b'c} = 20$ pF

$f_T = 150$ KHz

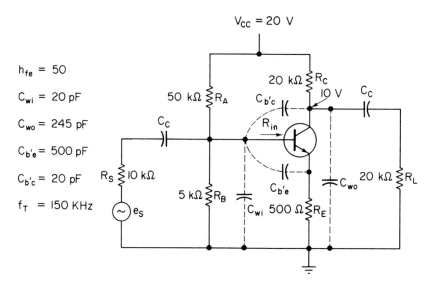

FIG. 8-15. Circuit for Example 8-5

input effects is

$$f_2 \simeq \frac{f_T}{h_{fe}} = \frac{150 \text{ kHz}}{50} = 3 \text{ kHz}$$

without considering feedback effects.

$$G_v = \frac{20 \text{ k}\Omega \,\|\, 20 \text{ k}\Omega}{h_{ib}} \simeq \frac{10 \text{ k}\Omega}{52 \, \Omega} \simeq 200$$

without feedback and $B = -R_E/R_L = -\frac{1}{10}$. Therefore, by Eq. (8-11),

$$f_{2fb} = 3 \text{ kHz}[1 - (\tfrac{1}{10} \times 200)] = 3 \text{ kHz} \times 21 = 63 \text{ kHz}$$

when feedback effects are taken into account.

2. Since $C_{b'c}$ and $C_{b'e}$ are given, we can double-check the previous result by calculating C_{shunt} as

$$C_{\text{shunt}} = C_{wi} + C_{b'e} + (1 + G_v)C_{b'c} = 20 \text{ pF} + 500 \text{ pF} + \left(1 + \frac{10 \text{ k}\Omega}{500 \, \Omega}\right) 20 \text{ pF}$$
$$= 940 \text{ pF}$$

and R_{shunt} as

$$R_{\text{shunt}} = R_S \,\|\, R_A \,\|\, R_B \,\|\, R_{IN} = 10 \text{ k}\Omega \,\|\, 50 \text{ k}\Omega \,\|\, 5 \text{ k}\Omega \,\|\, (h'_{fe} \times 500 \, \Omega)$$
$$\simeq 2.8 \text{ k}\Omega$$

Therefore, f_2 is

$$f_2 = 1/2\pi RC = 1/2\pi \times 2.8 \text{ k}\Omega \times 940 \text{ pF}$$
$$= 60 \text{ kHz}$$

This solution for f_2 is usually somewhat lower than the simpler solution using f_T, and it must be remembered that the first solution did not include the effects of current reduction from the imperfect current source ($R_S = 10$ k$\Omega \neq \infty$) or the stray input capacitance, C_{wi}.

3. Now solving for the output circuit high-frequency cutoff, we obtain from Eq. (8-13)

$$f_2 = \frac{1}{2\pi(20 \text{ pF} + 245 \text{ pF})(20 \text{ k}\Omega \| 20 \text{ k}\Omega)}$$
$$= \frac{1}{2\pi(265 \text{ pF})(10 \text{ k}\Omega)}$$
$$= 60 \text{ kHz}$$

4. Since the cutoff frequencies for input and output effects are equal, the overall value of f_2 from Eq. (8-14) is

$$f_2 \simeq \frac{60 \text{ kHz}}{(1.1)^2} = 50 \text{ kHz}$$

The high-frequency cutoff of CB and CC stages is usually an order of magnitude (or more) above those of CE stages, with CCs being even better than the CB amplifiers. Since any amplifier invariably consists of at least one CE stage, the exact f_2 of the CB or CC stages is usually immaterial, since the CE stage usually determines the high-frequency cutoff. However, f_2 for a CB stage can be approximated as being equal to the cutoff predicted by $C_{b'c}$ and R_L or f_α, where $f_\alpha \simeq 1.2 f_T$, whichever is lowest. The value of f_2 for a CC stage can be approximated as

$$f_{2CC} = \frac{1}{2\pi R_{\text{shunt}}(C_{b'c} + C_{b'e}/h_{fe})} \tag{8-15}$$

where R_{shunt} is the resistance seen from the input to ground. Notice the reduction of the effective value of $C_{b'e}$ by the factor $1/h_{fe}$ because of the 100 per cent in-phase feedback effects between the $C_{b'e}$ terminals. It is for this reason that CC stages have such good frequency response.

8-7 Tuned Amplifiers—Introduction

We have thus far discussed the factors that determine f_1 and f_2 for an amplifier with the idea that an f_1 as low as possible and an f_2 as high as possible were the ultimate goal. In a tuned amplifier, however, it is the goal to have a narrow bandwidth ($f_2 - f_1$) so that just a very specific range of frequencies will be amplified. This is necessary in a radio receiver, for instance, so that just one station is received at any one time. In addition, since electrical noise (unwanted electrical disturbances) is equally spaced throughout the radio frequency (RF) spectrum, selection of a narrow frequency slice in a tuned amplifier will reject the great majority of noise.

Figure 8-16 shows two simple *RLC* circuits—often termed *tuned circuits, tank circuits,* or *resonant circuits.* Tuned circuits are the basic building

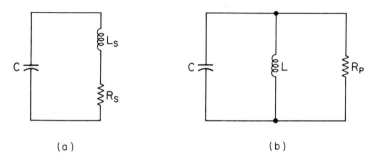

FIG. 8-16. (a) Tuned circuit—series resistance; (b) tuned circuit—parallel resistance

blocks of tuned amplifiers and hence this introduction to their basic principles. For practical purposes the capacitor can be considered as ideal (no leakage current), but the inductor's winding resistance has a significant effect and must be considered, as shown in Fig. 8-16a. Figure 8-16b shows the series winding resistance R_S as an equivalent parallel resistance R_P.

The *quality factor Q* of an inductor is a measure of its inductive reactance (at a specific frequency) to its dc winding resistance R_S:

$$Q_{\text{series}} = Q_s = \frac{\omega L}{R_S} \qquad (8\text{-}16)$$

For the equivalent parallel *LCR* circuit, Q_P is

$$Q_P = \frac{R_P}{\omega L} \qquad (8\text{-}17)$$

and conversion from the parallel to series resistive circuits is possible by the following formula:

$$R_P \simeq R_S Q^2 \qquad (8\text{-}18)$$

since $L_P = L_S$ and $Q_P = Q_S$ when Q is greater than 10—the usual case.

Figure 8-17 shows the impedance and phase angle of a parallel resonant circuit as a function of the resonant frequency f_o. The resonant frequency f_o is

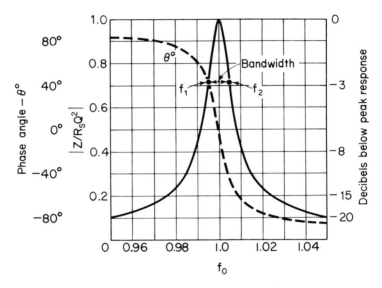

FIG. 8-17. Impedance and phase angle of a parallel resonant circuit versus frequency (From John D. Ryder, *Electronic Fundamentals and Applications*, 4th ed., Prentice-Hall, Inc., Englewood Cliffs, N.J., 1970.)

about equal to the frequency at which $X_C = X_L$ and can therefore be calculated as

$$X_C = X_L \quad \text{at } f_o$$

$$\frac{1}{2\pi f_o C} = 2\pi f_o L$$

$$f_o^2 = \frac{1}{(2\pi)^2 LC}$$

$$f_o = \frac{1}{2\pi\sqrt{LC}} \tag{8-19}$$

At resonance the impedance of the circuit reaches a maximum and is

$$Z_o = R_s Q^2 = \frac{L_S}{CR_S} \tag{8-20}$$

where Q_o is the quality factor of the inductor at resonance. The latter relationship of Eq. (8-20) indicates that Z_o is a function of the L/C ratio. Note that the circuit's impedance rises sharply to a maximum value at resonance. The steepness of this rise is determined by Q, with high Qs providing steep slopes. The bandwidth of the tank circuit is the range of frequencies between the points where the circuit impedance has fallen to 0.707 of its peak value.

The bandwidth (BW) is given by

$$\text{BW} = f_2 - f_1 = \frac{f_o}{Q} \qquad (8\text{-}21)$$

It is thus seen that a high-Q circuit will provide a narrow BW. If such a circuit is used as the load of a CE or common-source amplifier, the BW of this amplifier will be the same as the BW of the tuned circuit, since the voltage gain is directly proportional to load impedance. This, then, is the very basis of a tuned amplifier. The BW of a tuned circuit can also be shown to be

$$\text{BW} = \frac{1}{2\pi R_p C} \qquad (8\text{-}22)$$

This shows the dependence of BW on the capacitor size and is often a useful relationship.

EXAMPLE 8-6

Determine Q, f_o, Z_o, A_v, and the BW for the single-stage tuned amplifier of Fig. 8-18. The resistance, R_P, includes the effects of the inductor's dc winding resistance and the effect of the load impedance R_L shown in dashed lines in the figure.

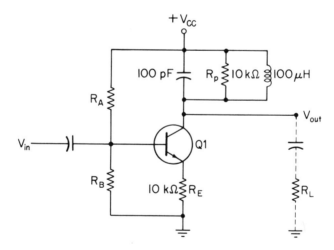

FIG. 8-18. Single-stage tuned amplifier

Solution:
From Eq. (8-19),

$$f_o = \frac{1}{2\pi\sqrt{LC}}$$

$$= \frac{1}{2\pi\sqrt{(100\ \mu\text{H})(100\ \text{pF})}}$$

$$= 1.59\ \text{MHz}$$

and from Eq. (8-17) then

$$Q_o = \frac{R_P}{\omega L} = \frac{10 \text{ k}\Omega}{1.59 \times 10^6 \times 2\pi \times 100 \times 10^{-6}}$$
$$= 10$$

The impedance at f_o is calculated from Eq. (8-20) as

$$Z_o = R_S Q_o^2$$

where $R_S \simeq R_P/Q^2$ from Eq. (8-18):

$$R_S = \frac{10 \text{ k}\Omega}{10^2} = 100 \text{ }\Omega$$

Therefore, the impedance of the tuned circuit at resonance is given by Eq. (8-20) as

$$Z_o = 100 \times Q^2$$
$$= 100 \text{ k}\Omega$$

We can now calculate the voltage gain at resonance as

$$G_v = \frac{R_C}{R_E} = \frac{Z_o}{R_E} = \frac{100 \text{ k}\Omega}{10 \text{ k}\Omega} = 10$$

with the BW between the 3 dB down limits of

$$\text{BW} = \frac{f_o}{Q} = \frac{1.59 \text{ MHz}}{10}$$
$$= 159 \text{ kHz}$$

from Eq. (8-21).

8-8 Tuned Amplifiers

More often than not, tuned amplifiers contain more than one stage. In these cases, the coupling between stages is usually via transformers, since the primary and secondary offer ready-made inductors, and dc levels between stages are thereby isolated without the need for coupling capacitors. Figure 8-19 shows such a situation in which the transformer $T1$ is shown to have a variable core. These tuning cores have threaded slugs that may be adjusted to provide variable core permeability and hence variable inductance. The inductance of an inductor is determined by

$$L = K\mu N^2 \qquad (8-23)$$

where K is a constant determined by inductor geometry, μ is the permeability of the core, and N is the number of turns of wire. Thus adjustment of a tuned

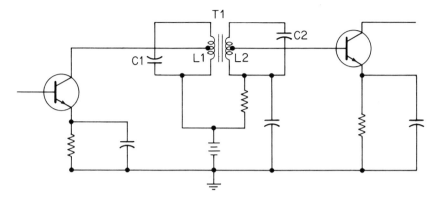

FIG. 8-19. Two-stage doubly tuned amplifier

circuit's f_o is possible. These transformers are visible as small square-shaped cans with a screw adjustment in virtually all today's standard radio receivers.

Notice the use of the tapped inductor configuration in Fig. 8-19. The use of tapped inductors in tuned amplifiers provides remarkable impedance transformations, as well as allowing the use of practicably available inductors and capacitors for many applications. Consider an application that calls for a 10 kHz BW centered at 455 kHz. This is the requirement of the intermediate frequency (IF) amplifiers of a standard AM radio receiver. From Eq. (8-22),

$$C = \frac{1}{2\pi \times 10 \text{ kHz} \times 10^3}$$
$$= 15{,}900 \text{ pF}$$

and then L is calculated from Eq. (8-19) as

$$L = \frac{1}{[(2\pi)455 \text{ kHz})]^2 \times 15{,}900 \text{ pF}}$$
$$= 7.7 \text{ microhenrys } (\mu\text{H})$$

Thus the size of the required capacitor of 15,900 pF is too large for the low-loss-type mica or ceramic devices available, and the inductor size of 7.7 μH is too small to be practicably built. Selection of the proper tapped inductor will solve the problem because of the inductive and capacitive transformation capabilities, as seen in the following two formulas:

$$L_1 = L\left(\frac{N_1}{N_2}\right)^2 \tag{8-24}$$

$$C_1 = C\left(\frac{N_2}{N_1}\right)^2 \tag{8-25}$$

where L_1 and C_1 are the effective value of L and C reflected through the use of an inductor with a total of N_1 turns and tapped N_2 turns from one end. For instance, in our example let us say that a 200 μH inductor is commercially available; the required turns ratio would be

$$\left(\frac{N_1}{N_2}\right)^2 = \frac{200 \ \mu\text{H}}{7.7 \ \mu\text{H}} = 26$$

Therefore

$$\frac{N_1}{N_2} = 5.1$$

Since the reflected value of capacitance must be 15,900 pF, we can solve for the value we must use from Eq. (8-25) as

$$C_1 = C\left(\frac{N_2}{N_1}\right)^2 = \frac{C}{(N_1/N_2)^2} = \frac{15{,}900 \ \text{pF}}{(5.1)^2}$$
$$= 610 \ \text{pF}$$

Therefore, the circuit of Fig. 8-20a will result in the required value of resonant frequency and BW. Figure 8-20b shows the circuit that it is equivalent to. The circuit in part a of the figure is practicably obtainable; the one in part b is not.

In practice the preceding example would be only one stage of a whole series of tuned circuits in an IF amplifier. The inductor of this example would be the primary of an IF transformer, and the secondary would form the input tank to a following stage. The bandwidth of n identical (in BW) tuned stages is given by

$$\text{BW}_{n \text{ stages}} = \text{BW}(2^{1/n} - 1)^{1/2} \qquad (8\text{-}26)$$

where BW is the bandwidth of the n individual stages.

EXAMPLE 8-7

If four 10 kHz BW tuned circuits are cascaded, determine the resultant overall bandwidth.

Solution:

$$\begin{aligned}\text{BW}_{4 \text{ stages}} &= \text{BW}(2^{1/4} - 1)^{1/2} \\ &= 10 \ \text{kHz}(1.19 - 1)^{1/2} \\ &= 4.35 \ \text{kHz}\end{aligned}$$

If two or more tuned circuits are cascaded, it is possible to obtain a more ideal response if each tank circuit is tuned to a slightly different frequency. This technique, known as *stagger tuning*, is illustrated in Fig. 8-21 for a two-

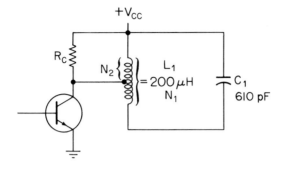

(a) practical version

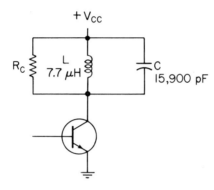

(b) equivalent but unpractical version

FIG. 8-20. 455-kHz, 10-kHz BW tuned amplifier

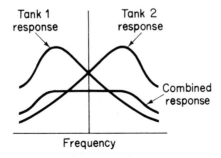

FIG. 8-21. Response of two-stage stagger tuning

stage circuit. Note that the L_1C_1 tank is tuned to a lower frequency than L_2C_2, and an increased BW with a flat passband and relatively steep sides results from this configuration. Up to four stages are used, and the more stages used the flatter will be the passband and the steeper will be the falloff. Note also the loss of gain that results with increased bandwidth. When optimum stagger tuning is utilized, it is said to be a *Butterworth* arrangement. This optimum arrangement is the closest thing to a rectangular response curve for the filter.

There are many practical applications for the use of tuned power amplifiers—notably in radio transmitters. In such an amplifier, where thousands of watts may be involved, class C operation is often used, since it offers greater efficiency than even class B. In class C operation, a reverse bias causes the transistor (or tube in many high-power situations) to be cut off for more than 180° of its input cycle. The active device then acts as a switch to supply energy to its *LC* tank load during a relatively short duty cycle. The output is then a sine wave, because of the flywheel effect in the tank, with the active device supplying enough energy to make up for the power being drawn to the load. Some radio signals require a transmitter output that is proportional to the input, in which case a Class B tuned power amplifier is utilized; such an amplifier is usually referred to as a *linear amplifier*.

PROBLEMS

1. Express the following in decibels.
   (a) $G_v = 85$ (b) $V_o/V_i = 0.16$
   (c) $G_P = 100$

2. (a) Determine f_1 for a high pass RC circuit where $R = 10\,k\Omega$ and $C = 5\,\mu F$.
   (b) Reverse the position of the resistor and capacitor and determine f_2.

3. Repeat Problem 2 with R decreased by a factor of 10 and with both R and C reduced by a factor of 10.

4. How many decibels down will the circuit of Problem 2a be at 100 Hz? 20 Hz?

5. Determine f_1 for the amplifier of Fig. P8-5.

6. What would be the easiest way to increase f_1 for Problem 5?

7. Determine f_1 for the amplifier of Fig. P8-7.

8. Determine f_2 for the amplifier of Fig. P8-7 if $C_{gs} = 5\,pF$ and $C_{gd} = 5\,pF$ and $g_{fs} = 2000\,\mu\mho$.

9. Determine f_2 for the amplifier of Fig. P8-7 if the 1 $k\Omega$ source resistor were completely bypassed and $g_{fs} = 5000\,\mu\mho$.

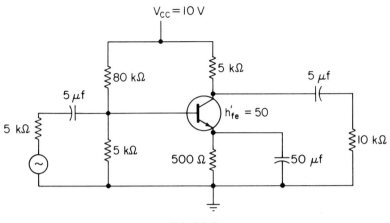

FIG. P8-5.

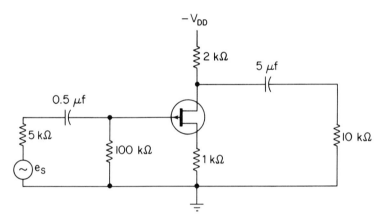

FIG. P8-7.

10. If two stages as shown in Fig. 8-14 were cascaded without having an emitter bypass, what would f_2 be due to input shunt capacitances only, and what would be the overall voltage gain? How does this compare to the same results for the single-stage version with emitter bypass?

11. The circuit of Fig. 8-13 is to be completely analyzed for its high-frequency response. It has the following values: $R_S = 15\ \text{k}\Omega$, $R_A = 60\ \text{k}\Omega$, $R_B = 6\ \text{k}\Omega$, $R_C = 30\ \text{k}\Omega$, $R_L = 30\ \text{k}\Omega$, $C_{b'c} = 5\ \text{pF}$, $C_{b'e} = 25\ \text{pF}$, $R_E = 2\ \text{k}\Omega$, $C_{wi} = 20\ \text{pF}$, $C_{wo} = 200\ \text{pF}$, C_E is omitted, $h'_{fe} = 100$, $V_{cc} = 15\ \text{V}$, and $V_c = 7.5\ \text{V}$. Analyze all aspects of its high-frequency response.

12. What effect on f_2 results when $C_E = 90\ \mu\text{F}$ in Problem 11?

13. The transistor of Problem 12 is to be used as a CC stage as shown in Fig. P8-13. Determine f_2 for this stage.

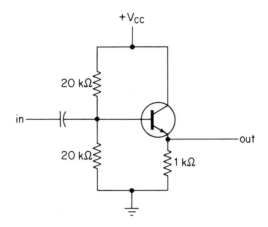

FIG. P8-13.

14. A parallel LC tank circuit has $C = 0.01\ \mu F$ and $L = 0.5$ mH and $R_S = 20\ \Omega$. Determine
(a) f_o (b) Q
(c) Z_o (d) BW

15. Design a single-stage tuned amplifier to fulfill the following requirements:
(a) $f_o = 225$ kHz   (b) BW $= 8$ kHz
The collector load resistance is 1 kΩ and a 200 μH inductor must be used. Determine where to tap it for proper operation and the proper capacitor to use.

16. If three identical stages as designed in Problem 15 were cascaded, calculate the resultant bandwidth.

9

Multistage Amplifiers

9-1 Introduction

Multistage amplifiers are nothing more than a combination of two or more active amplifying devices connected to give more gain than one of them could. Most amplifiers have more than one stage, and hence usually electronic personnel must deal with more than just a single-stage amplifier.

Unfortunately, the analysis of multistage amplifiers is not often an easy task for the student. To be successful he must be able to logically apply the principles developed in the previous chapters in a very careful fashion. There are often many interactions between stages that require a certain amount of experience to allow successful results. This is true whether the job is to design, explain, repair, or apply the multistage amplifier.

In a subtle way you have already been introduced to some simple multistage amplifiers in the previous chapters. This was done to facilitate explanation of effects and applications in those chapters. It should also serve to ease the transition into this study. The major goal here is to provide enough experience with some common multistage amplifiers to enable you to grapple with whatever configurations you may run into. This experience should include accompanying laboratory work, because only through direct confrontation can most students obtain a realization of the total problem involved.

The number of different multistage amplifiers is countless. Those presented here are the most common forms, but obviously the possible variations are limited only by the designer's imagination.

9-2 Darlington Compound

A *Darlington compound* is a connection of two transistors that act as a single transistor with an effective h'_{fe} equal to the product of the h'_{fe} of each individual transistor. The most common form is shown in Fig. 9-1. The

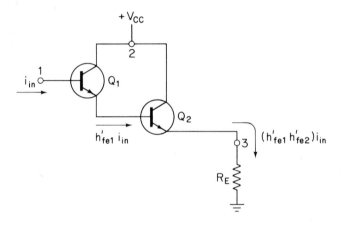

FIG. 9-1. Darlington compound

input current to the base of $Q1$ is amplified by a factor of h'_{fe1} in the emitter of $Q1$. This emitter current of the first stage is the base current of the second stage. It gets amplified by a factor of h'_{fe2} in the load resistor. Hence $G_i = h'_{fe1} h'_{fe2}$ and if identical transistors are used will simply equal h'^2_{fe}.

The Darlington pair acts as a single transistor with points 1, 2, and 3 of Fig. 9-1 equal to the base, collector, and emitter, respectively. They are commercially available in a single package with only those three leads brought out. This multistage package makes it possible to offer a much higher effective h_{fe} than is normally available with a single transistor. The input impedance is also of interest to us. Looking into the base of $Q2$ we "see" $h'_{fe2} \times R_E$ ohms. But this input impedance is the load impedance of stage 1, so looking into the base of $Q1$ we "see" $h'_{fe1}(h'_{fe2}R_E)$ Ω, or simply $h'^2_{fe}R_E$ if identical transistors are used. The Darlington compound is certainly capable of high input impedances. The impedance transformation capabilities of this circuit are enormous, as illustrated by the following example.

EXAMPLE 9-1

A CE amplifier stage is to drive a 100 Ω load to a 10 V p-p level. A 1 V p-p input signal is available. First determine the resultant output if the 100 Ω load is capacitively coupled directly to the CE stage, as shown in Fig. 9-2a, and then redetermine the output when a Darlington pair is used as a "buffer" between Q1 and the load, as in Fig. 9-2b.

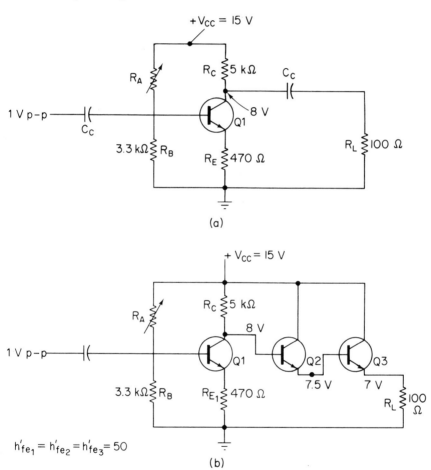

FIG. 9-2. (a) CE amplifier; (b) CE Darlington compound

Solution:

The voltage gain of the CE amplifier in Fig. 9-2a is given by the approximation R_L/R_E, where R_L equals $R_C \| R_L$:

$$G_v \simeq \frac{100\,\Omega \,\|\, 5\text{ k}\Omega}{470\,\Omega} \simeq \frac{1}{5}$$

It is seen that attenuation has occurred due to overloading the stage, and, by itself, the stage does not have enough current gain to drive that heavy a load. The circuit of Fig. 9-2b will provide a load to stage 1 equal to 5 k$\Omega \parallel Z_{\text{IN Darlington}}$:

$$R_{L1} = R_{C1} \parallel (h_{fe}'^2 \times R_L)$$
$$= 5 \text{ k}\Omega \parallel (50^2 \times 100 \text{ }\Omega)$$
$$= 5 \text{ k}\Omega \parallel 250 \text{ k}\Omega$$
$$\simeq 5 \text{ k}\Omega$$

Hence G_{v1} is

$$G_{v1} \simeq \frac{5 \text{ k}\Omega}{470 \text{ }\Omega} \simeq 10$$

and the voltage gain of the Darlington is about 1, since each CC stage has a G_v of about 1. Therefore, G_{voa} is 10 and the design goal of a 10 V p-p output has been met.

Notice the convenient ability of direct coupling between the CE output and the two following CC stages in Fig. 9-2b. The dc output (collector voltage) of the CE stage should be about half the supply voltage, as should the input of a CC stage when large-signal amplification is taking place. This works out to the approximate dc levels shown in the figure and does not allow the 10 V p-p signal at $Q1$'s collector to be clipped at any ensuing point in the amplifier.

Whenever a load causes a severe loss in voltage gain (loading effect), it is the usual practice to step up that load impedance via a single CC stage, or a double CC stage (Darlington pair) if an even greater impedance transformation is necessary. In some instances a Darlington pair may be used as the first stage of an amplifier system when working from a very high impedance source. Consider a transistor with an output voltage of 5 mV p-p and an internal resistance of 200 kΩ. It is normally not possible to obtain an input impedance (Z_{in}) of 200 kΩ with a CE stage, and hence severe attenuation of the already small signal would result, as illustrated in Fig. 9-3. It is seen from the figure that the already small input signal (5 mV) has gotten even smaller as a result of loading. These very small signals will be susceptible to noise pickup, and hence a much better solution would be as shown in Fig. 9-4. The minimum input impedance is about equal to the 20 MΩ bias resistor in parallel with $h_{fe}'^2(1 \text{ k}\Omega \parallel 3 \text{ k}\Omega)$. Therefore, assuming a gain of 1 through the Darlington, the amount of signal reaching the base of $Q3$ will be about 4.5 mV, as shown, and its gain of 10 results in a 45 mV output signal with a 3 kΩ output impedance.

EXAMPLE 9-2

Determine the range of bias voltage at the emitter of $Q2$ in Fig. 9-4 if h_{fe}' can vary from 50–100 for $Q1$ and $Q2$.

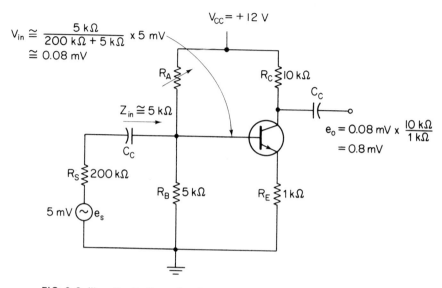

FIG. 9-3. "Loading" effect of typical CE stage on a high impedance source

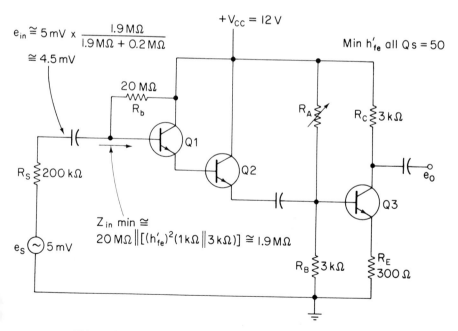

FIG. 9-4. Darlington and CE driven by high impedance source

Solution:
If we assume a 0.5 V drop for each base emitter junction, then

$$I_{E2} \times R_{E2} + I_{B1} \times R_{B1} = 11 \text{ V}$$
$$(50^2 I_{B1})1 \text{ k}\Omega + I_{B1}(20 \text{ mA}) = 11 \text{ V} \quad \text{for minimum } h'_{fe}$$
$$(100^2 I_{B1})1 \text{ k}\Omega + I_{B1}(20 \text{ mA}) = 11 \text{ V} \quad \text{for maximum } h'_{fe}$$

Solving these two equations results in an I_{B1} of about 0.5 µA when $h'_{fe} = 50$ and an I_{B1} of about 0.365 µA when $h'_{fe} = 100$. Thus the dc current through the 1 kΩ resistor will vary from

$$V_{E2\min} = 1.25 \text{ mA} \times 1 \text{ k}\Omega = 1.25 \text{ V}$$
to $\quad V_{E2\max} = 3.65 \text{ mA} \times 1 \text{ k}\Omega = 3.65 \text{ V}$

These are acceptable bias voltages for an ac signal of this low level (4.5 mV).

9-3 FET–BJT Combinations

The ability to work from high impedance sources is a trait of FETs as well as the Darlington pairs of the previous section. For instance, the performance of the amplifier of Fig. 9-4 is essentially duplicated by the circuit in Fig. 9-5. The voltage gain of the source follower is almost 1, and a G_v of 10 for stage 2 yields an output of 50 mV. The input impedance of the FET is about 10 MΩ, and hence it presents negligible loading to the source. Zero-bias drain current (I_{DSS}) ranges of about 1–10 mA are allowable, resulting in source Q-point voltages of 1–10 V. This circuit is more compact than the Darlington version, and the choice between the two would probably boil down to an economic one.

Multistage amplifiers often include such a combination of FETs and BJTs as shown in Fig. 9-5. Obviously, the major advantage of FETs is the ability to work from a high impedance source, but also of importance is the ability to use *high interstage impedance* levels. Thus coupling capacitors may be smaller for a given low-frequency requirement, and level-control (and frequency-response-control) potentiometers may be of very low power ratings. Thus FET benefits added to the high-voltage-gain capabilities of BJTs make their combination an attractive arrangement. The extreme simplicity in cascading a number of enhancement-mode MOSFET stages was demonstrated in Section 6-5.

Figure 9-6 illustrates another FET–BJT amplifier, which shows some of the advantages of their combination. The amplifier is very stable to temperature changes and has a highly predictable voltage gain. It has a high input

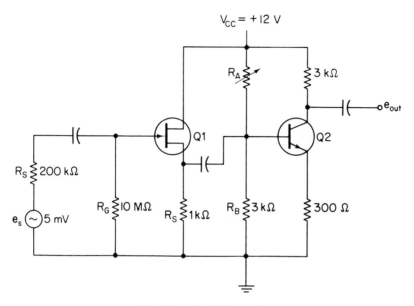

FIG. 9-5. FET–BJT amplifier

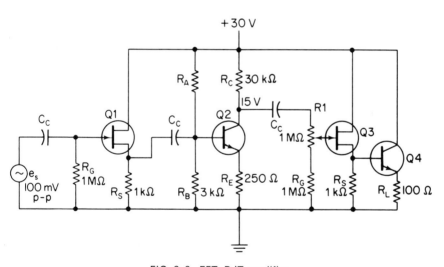

FIG. 9-6. FET–BJT amplifier

impedance because of the FET input stage $Q1$, and high voltage gain because of the BJT $Q2$, with a high value of R_C. Normally, driving from a high value of R_C causes loading problems and hence negation of the high voltage gain, but the driven stage in this case, $Q3$, is another FET with high input impedance. Thus no loading takes place, and the voltage gain is maintained.

Adjustment of voltage gain is allowed with a very low power potentiometer, $R3$, since power is equal to V^2/R and R is very large in this case.

EXAMPLE 9-3

Determine the ac load voltage range for the amplifier of Fig. 9-6 given the following information:

$$Q1 - I_{DSS} = 2\text{–}7 \text{ mA}, \qquad Q2 - h'_{fe} = 50\text{–}100$$
$$g_{fs} = 3000\text{–}7000 \text{ }\mu\mho$$
$$Q3 - I_{DSS} = 8\text{–}15 \text{ mA}, \qquad Q4 - h'_{fe} = 50\text{–}100$$
$$g_{fs} = 2000\text{–}8000 \text{ }\mu\mho$$

Solution:
The voltage gain of the first stage is about equal to 1:

$$G_{v1} \simeq 1$$
$$G_{v2} \simeq \frac{R_{\text{collector}}}{h_{ib} + R_E} = \frac{30 \text{ k}\Omega \parallel R_{\text{IN3}}}{52 \text{ }\Omega + 250 \text{ }\Omega} \simeq \frac{30 \text{ k}\Omega \parallel 2 \text{ M}\Omega}{300 \text{ }\Omega} \simeq \frac{30 \text{ k}\Omega}{300 \text{ }\Omega}$$
$$G_{v2} \simeq 100$$

Stages 3 and 4 have voltage gains of about 1 since they are both follower stages. Therefore, the overall voltage gain is

$$G_{voa} = G_{v1} \times G_{v2} \times G_{v3} \times G_{v4}$$
$$\simeq 1 \times 100 \times 1 \times 1 = 100$$

Notice that the approximate gain did not depend on the active-device characteristics in any way, but was determined by resistor values and is therefore highly predictable. The output voltage then will be

$$e_{\text{out}} = G_{voa} \times e_{\text{in}}$$
$$= 100 \times 100 \text{ mV p-p}$$
$$= 10 \text{ V p-p}$$

This output could be cut in half by the voltage divider action of the 1 MΩ potentiometer, $R1$, and the 1 MΩ gate resistor, R_G.

EXAMPLE 9-4

Determine the proper value of R_A in Fig. 9-6 for a 15 V Q-point at $Q2$'s collector when $h'_{fe2} = 75$. If h'_{fe2} ranges from 50–100, will this have any appreciable effect on the Q-point?

Solution:
If $V_{C2} = 15$ V, then $I_{C2} = 15$ V/300 kΩ = 0.5 mA. Then looking into the base of $Q2$, $h'_{fe2}(h_{ib2} + 250 \text{ }\Omega)$ ohms is seen. Since $I_{E2} \simeq I_{C2} = 0.5$ mA, $h_{ib} = 52 \text{ }\Omega$ by Shockley's relation and $75 \times (52 + 250) = 22.5$ kΩ is seen. Since the base current

must be $0.5 \text{ mA}/h_{fe2} = 6\frac{2}{3} \mu\text{A}$, $R1$ must supply that plus current for $R2$. The current through $R2$ will be $(V_{E2} + V_{BE})/R2$ or

$$I_{R2} \simeq \frac{0.5 \text{ mA}(250 \text{ }\Omega) + 0.7 \text{ V}}{3 \text{ k}\Omega}$$

$$\simeq \frac{0.825 \text{ V}}{3 \text{ k}\Omega} \simeq 0.275 \text{ mA}$$

$$\simeq 275 \text{ }\mu\text{A}$$

The voltage across $R1$ will be $30 \text{ V} - 0.825 \text{ V}$ or about 29.2 V, and since it must supply $275 \text{ }\mu\text{A} + 6\frac{2}{3} \text{ }\mu\text{A}$, it must equal

$$R_1 \simeq \frac{29.2 \text{ V}}{282 \text{ }\mu\text{A}} \simeq 100 \text{ k}\Omega$$

Since the base current is so small compared to I_{R2}, it is safe to assume that changes in it due to h_{fe} variations will have negligible effects on the Q-point.

9-4 Multistage Feedback

For the most part we have dealt with feedback within one stage, and this is known as *local feedback*. Multistage amplifiers offer a means of introducing feedback to an entire amplifier. A two-stage amplifier that we shall use to study this effect is shown in Fig. 9-7. Notice that the output signal is fed back to the emitter of $Q1$ through the feedback resistor R_F. Since the current thus fed back is in phase with the emitter current of $Q1$, it will be additive through R_{E1}. This means that for a given V_{in}, the ac V_{be} of $Q1$ will be reduced by this fed-back signal by Kirchhoff's voltage law around the V_{in}, V_{be1}, and R_{E1} loop. With less ac voltage across $Q1$'s base–emitter junction, less will appear at its collector, and negative feedback has occurred. The effects of enough negative feedback are to

1. Reduce the gain to a stable value predicted by resistor values.
2. Increase the amplifier's input impedance.
3. Improve the frequency response at both the high and low ends of the frequency spectrum.

This multistage feedback could be accomplished by feeding back the output signal to the first-stage emitter resistor of any even number of stages. To do so with an odd number of stages would introduce positive feedback, which we shall discuss in Chapter 10.

To determine the gain–impedance characteristics of this amplifier, we shall start with the general feedback equation [(Eq. 5-4)]. It provides us the voltage gain with feedback G'_v as a function of the nonfeedback voltage gain

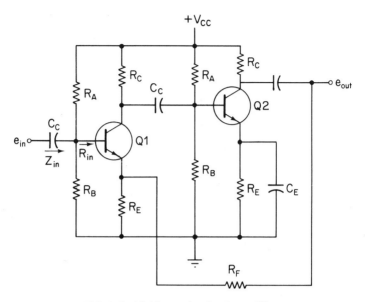

FIG. 9-7. Multistage feedback amplifier

G_v and the feedback factor B:

$$G'_v = \frac{G_v}{1 - BG_v} \tag{5-4}$$

For the amplifier of Fig. 9-7, G_v is equal to the product of nonfeedback gains G_{v1} and G_{v2}, and B is equal to

$$B \cong -\frac{R_{E1}}{R_F} \tag{9-1}$$

The ideal situation is to make G_v very large and R_F/R_E large but at least 10 times smaller than G_v. In that case the voltage gain of this amplifier is approximately

$$G'_v \cong \frac{R_F}{R_E} \tag{9-2}$$

as will be shown in the following example.

EXAMPLE 9-5

Suppose that the amplifier of Fig. 9-7 has the following values:

$$R_{C1} = R_{C2} = 10 \text{ k}\Omega$$
$$R_{E1} = 100 \text{ }\Omega, \quad R_{B2} = 5 \text{ k}\Omega$$
$$R_F = 10 \text{ k}\Omega$$

$$V_{CC} = 20 \text{ V}, \quad h'_{fe1} = h'_{fe2} = 100$$
$$V_{C2} = 10 \text{ V}$$

Determine the voltage gains without and with feedback, G_v and G'_v, respectively.

Solution:

$$G_v = G_{v1} \times G_{v2}$$

$$G_{v1} = \frac{10 \text{ k}\Omega \| Z_{in2}}{100 \text{ }\Omega} \simeq \frac{10 \text{ k}\Omega \| (5 \text{ k}\Omega \| h'_{fe2} h_{ib2})}{100 \text{ }\Omega} \simeq \frac{1.46 \text{ k}\Omega}{100 \text{ }\Omega} = 14.6$$

$$G_{v2} = \frac{R_L}{h_{ib}} = \frac{10 \text{ k}\Omega}{26 \text{ }\Omega} \simeq 400$$

Therefore,

$$G_v = 14.6 \times 400 = 5820$$

The feedback factor B is

$$B \simeq -\frac{R_E}{R_F} = -\frac{100 \text{ }\Omega}{10 \text{ k}\Omega} = -\frac{1}{100}$$

Therefore, G'_v is

$$G'_v = \frac{G_v}{1 - BG_v} = \frac{5820}{1 - [-(1/100)](5820)}$$
$$= \frac{5820}{1 + 58.2} \simeq 100$$

The approximate formula, Eq. (9-2), yields the same result:

$$G'_v \simeq \frac{R_F}{R_E} = \frac{10 \text{ k}\Omega}{100 \text{ k}\Omega} = 100$$

and thus the voltage gain with feedback is in fact predicted by resistor values—a desirable feature.

By derivation, it can be shown that the input impedance looking into the base of $Q1$ of this form of amplifier is approximated by

$$R_{IN} \simeq \frac{G_v}{G'_v} \times h'_{fe} \times h_{ib1}$$

The overall amplifier input impedance (Z_{in}) is then given by the parallel combination of R_{in} and the bias resistors for stage 1. This usually results in an input impedance predicted solely by bias resistors (see Example 9-6). Once the complete amplifier's input impedance is known, the amplifier's overall current gain can easily be determined with the aid of the TGIR, as shown in Example 9-6.

EXAMPLE 9-6

The amplifier of Example 9-5 has stage 1 bias resistors of 10 kΩ and 100 kΩ, and they result in a $Q1$ collector voltage of 10 V. Determine Z_{in} and the overall current gain G_{ioa}.

Solution:
Z_{in} is $R_{in} \| 100 \text{ k}\Omega \| 10 \text{ k}\Omega$:

$$R_{in} \simeq \frac{G_v}{G_v'} h'_{fe1} \times h_{ib1}$$

$$= \frac{5820}{100} \times 100 \times h_{ib2}$$

h_{ib2} is given by $0.026/(10 \text{ V}/10 \text{ k}\Omega) = 26 \text{ }\Omega$, and therefore

$$R_{in} = 5820 \times 26 \text{ }\Omega \simeq 152 \text{ k}\Omega$$

Therefore

$$Z_{in} = R_{in} \| R_1 \| R_2$$
$$= 152 \text{ k}\Omega \| 100 \text{ k}\Omega \| 10 \text{ k}\Omega$$
$$\simeq 8.6 \text{ k}\Omega$$

Using the TGIR to obtain the current gain,

$$G_{ioa} = G_{voa} \times \frac{Z_{in}}{R_L}$$

$$\simeq 100 \times \frac{8.6 \text{ k}\Omega}{10 \text{ k}\Omega}$$

$$= 86$$

9-5 Three-Stage Amplifier

The three-stage amplifier of Fig. 9-8 is a common everyday configuration. We shall make an analysis of this circuit using the techniques we have thus far developed. Since virtually all these techniques have involved the use of approximations, we shall verify our results by comparing them against a carefully performed computer solution and the results of actually constructing and measuring the circuit's performance. Notice that stages 1 and 2 are CE circuits and the third is a direct-coupled CC emitter follower. The first stage employs emitter feedback, whereas the second stage's emitter resistor is bypassed by C_{E2}.

The amplifier's overall input impedance (Z_{in}) is calculated approximately as $15 \text{ k}\Omega \| 82 \text{ k}\Omega \| h'_{fe1}(1 \text{ k}\Omega) = 12.1 \text{ k}\Omega$. The gain of the first stage, G_{v1},

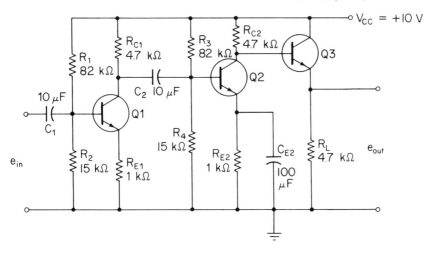

FIG. 9-8. Three-stage amplifier (Courtesy of McGraw-Hill Book Company.)

is simply R_{L1}/R_E using Eq. (5-9). The only catch here is to realize that R_L for Q1 is the 4.7 kΩ collector resistor in parallel with the input impedance of the second stage. Since the emitter of Q2 is bypassed,

$$\begin{aligned} Z_{IN2} &= 15 \text{ k}\Omega \,\|\, 82 \text{ k}\Omega \,\|\, (h'_{fe2} \times h_{ib2}) \\ &= 15 \text{ k}\Omega \,\|\, 82 \text{ k}\Omega \,\|\, (250 \times 26 \text{ }\Omega) \\ &= 15 \text{ k}\Omega \,\|\, 82 \text{ k}\Omega \,\|\, 6.25 \text{ k}\Omega \\ &= 4.18 \text{ k}\Omega \end{aligned}$$

Therefore, $R_{L1} = 4.7 \text{ k}\Omega \,\|\, 4.18 \text{ k}\Omega = 2.22 \text{ k}\Omega$, and we can calculate G_{v1} as

$$G_{v1} \simeq \frac{R_L}{R_E} = \frac{2.22 \text{ k}\Omega}{1 \text{ k}\Omega} = 2.22 \tag{5-9}$$

Since the second stage is bypassed, G_{v2} will equal R_{L2}/h_{ib2}. The load is 4.7 kΩ $\|\, Z_{in3}$. The input impedance of the third stage is $h'_{fe3} \times 4.7$ kΩ, or 250×4.7 kΩ, or 1.175 MΩ. Therefore, R_{L2} is 4.7 kΩ $\|\, 1.175$ MΩ $\simeq 4.7$ kΩ and

$$G_{v2} = \frac{4.7 \text{ k}\Omega}{26 \text{ }\Omega} = 181$$

Therefore, the amplifier's overall voltage gain, G_{voa}, can be calculated, assuming a G_v of 0.9 for the CC stage, as

$$\begin{aligned} G_{voa} &= G_{v1} G_{v2} G_{v3} \\ &= 2.22 \times 181 \times 0.9 \\ &= 360 \end{aligned}$$

The output impedance of the amplifier is 4.7 kΩ in parallel with the impedance seen looking into the emitter of $Q3$, which is $h_{ib3} + R_{C2}/h'_{fe3}$. Therefore, R_{out} is

$$R_{out} = \left(26 \text{ Ω} + \frac{4.7 \text{ kΩ}}{250 \text{ Ω}}\right) \| 4.7 \text{ kΩ} \simeq 26 \text{ Ω} + 19 \text{ Ω}$$
$$= 45 \text{ Ω}$$

Thus we now have the important gain–impedance relationships for this amplifier with an input impedance of 12.1 kΩ, an output impedance of 45 Ω, and a midband voltage gain of 360.

Let us now take a look at the amplifier's low-frequency characteristics. The two 10 μF coupling capacitors and the 100 μF bypass will be the only factors affecting this response. The first coupling capacitor forms a time constant with the amplifier's input impedance of 12.1 kΩ. Hence f_1 due to it is

$$f_1 = \frac{1}{2\pi RC} = \frac{1}{2\pi \times 12.1 \text{ kΩ} \times 10 \text{ μF}}$$
$$\simeq 1.3 \text{ Hz}$$

The coupling capacitor between $Q1$ and $Q2$ sees an impedance of 4.7 kΩ + (15 kΩ $\|$ 82 kΩ $\|$ $h'_{fe2} h_{ib2}$) ≃ 9.2 kΩ. Therefore, f_1 due to C_2 is

$$f_1 = \frac{1}{2\pi \times 10 \text{ μF} \times 9.2 \text{ kΩ}}$$
$$\simeq 1.7 \text{ Hz}$$

The bypass capacitor "sees" an impedance of 1 kΩ in parallel with h_{ib2} + (4.7 kΩ $\|$ 82 kΩ $\|$ 15 kΩ)/h'_{fe2}. This is

$$1 \text{ kΩ} \| (26 \text{ Ω} + 13.6 \text{ Ω}) \simeq 40 \text{ Ω}$$

Hence f_1 due to C_{E2} is

$$f_1 = \frac{1}{2\pi (40 \text{ Ω}) 100 \text{ μF}}$$
$$\simeq 40 \text{ Hz}$$

Thus the low-frequency response is determined almost solely by the bypass capacitor C_{E2} and the 3 dB down point for the overall amplifier is 40 Hz.

The high-frequency cutoff will be determined by stage 2. Stage 1 has a low voltage gain, and hence the Miller-effect capacity is minimized; stage 3, the CC emitter follower, has inherently extremely good frequency response. According to the procedure developed for high-frequency analysis in Chapter

8, we calculate f_2 on the input side of the circuit as being caused by (see Section 8-6)

$$C_{in} = C_{b'e} + (1 + G_{v2})C_{b'c}$$
$$= 139 \text{ pF} + (182)4 \text{ pF} = 967 \text{ pF}$$

That capacitance "sees" a resistance of 4.7 kΩ ‖ 15 kΩ ‖ 82 kΩ ‖ $h'_{fe2}h_{ib2}$ = 2.22 kΩ. Therefore, f_2 due to input effects is

$$f_2 = \frac{1}{2\pi RC} = \frac{1}{2\pi \times 2.22 \text{ kΩ} \times 967 \text{ pF}}$$
$$\simeq 83 \text{ kHz}$$

Then f_2 due to output effects is due to the time constant of $C_{b'c}$ and 4.7 kΩ ‖ $R_{in3} \simeq 4.7$ kΩ. Therefore,

$$f_{2out} = \frac{1}{2\pi \times 4 \text{ pF} \times 4.7 \text{ kΩ}} = 8.5 \text{ MHz}$$

Clearly, f_2 is caused by the input effects and is equal to 83 kHz. This completes the analysis of the amplifier. The results are summarized in Table 9-1, and the results of a computer analysis and laboratory tests on this circuit are also included. The validity of the approximate analysis developed in this book is obviously substantiated by these results.

Table 9-1 Amplifier Analysis Comparison

	APPROXIMATE ANALYSIS	COMPUTER* ANALYSIS	MEASURED*
G_{voa}	360	353	340
Z_{IN}	12.1 kΩ	12.0 kΩ	13 kΩ
R_{OUT}	45 Ω	47 Ω	53 Ω
f_1	40 Hz	35 Hz	33 Hz
f_2	83 kHz	90 kHz	81 kHz

*Courtesy of McGraw-Hill Book Company.

Surprisingly close agreement is shown between the three methods of analysis. The experimental results were obtained without using 2N3565s selected for h'_{fe}'s of 250 as used in the calculations. The range of h_{fe} for this transistor is 120–750, and resistors and capacitors had ±10 per cent tolerances. The computer analysis does not appear to provide any better results than the approximate solution. However, a computer program of a circuit allows a simplified worst-case analysis; i.e., it provides rapid analysis of amplifier performance under the worst possible combination of component characteristics. This is often done on circuits to be produced in large numbers to ensure adequate performance under the worst possible conditions.

9-6 Audio-Power-Amplifier Design Example

In this section a possible design for an audio amplifier will be presented. The specifications for the amplifier are as follows:

$$e_s = 50 \text{ mV p-p}$$
$$R_S = 5 \text{ k}\Omega$$
$$P_{out} = 10 \text{ W into 8 }\Omega \text{ speaker}$$
$$V_{CC} = 40 \text{ V}$$
all transistors $h'_{fe} = 100$
low-frequency cutoff $f_1 = 20 \text{ Hz}$

The first step is to determine the required output voltage. Since power equals V_L^2/R, we can solve for V_L:

$$V_L = \sqrt{PR} = \sqrt{10 \text{ W} \times 8 \text{ }\Omega}$$
$$= 8.95 \text{ V rms}$$

Hence V_L must be $8.95 \times 2 \times \sqrt{2} = 25.3$ V p-p. Then the overall voltage gain of the amplifier must be $25.3/0.050 \simeq 500$. We shall use a complementary symmetry push–pull design to obtain maximum efficiency and bias it class AB to eliminate crossover distortion. Recall the basic form of this amplifier from Section 7-8, as shown in Fig. 9-9. Shown in Fig. 9-10 is a CC driver stage serving as an input to Fig. 9-9. The driver is used to provide impedance transformation from the low input impedance of the push–pull transistors. It would seem logical to drive $Q3$ directly from a CE stage whose Q-point was adjusted to 21 V. A possible drive configuration for $Q3$ is shown in Fig. 9-10. It is these two stages, $Q5$ and $Q4$, that are used to provide voltage gain. Since $Q5$ has a bypassed emitter and $Q4$ does not, it seems likely that the G_v of 500 could be accomplished with the two stages. If not, however, an additional preamplifier may be added later.

To determine resistor values, we shall work from the output of the driver stage, $Q4$, back to the input. Looking into the base of $Q3$, an input impedance of $[h'_{fe3} \times ((R_{E3} || R_1 || R_2)|| \times h'_{fe1 \text{ or } 2} \times (R_L + R_E))]$ is seen. If R_{E1}

Audio-Power-Amplifier Design Example   217

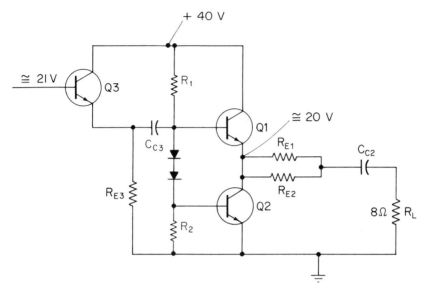

FIG. 9-9. Output stages for an audio amplifier

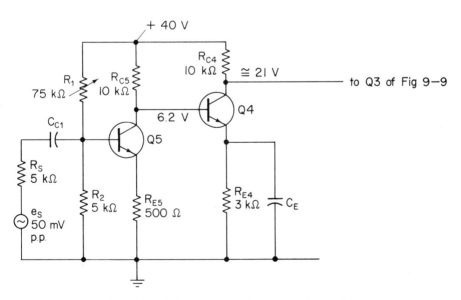

FIG. 9-10. Drive for output stage shown in Fig. 9-9

$= R_{E2} = 1\,\Omega$, and $R_{E3} = 1\,k\Omega$, this impedance would be $h'_{fe3}[1\,k\Omega\,(1\,k\Omega\,\|\,1\,k\Omega)\|\,h'_{fe1\,\text{or}\,2} \times (1\,\Omega + 8\,\Omega)]$ or $25\,k\Omega$. Thus R_{C4} could easily be $10\,k\Omega$ and $Q4$'s output would not be significantly loaded. Since $Q4$'s collector voltage should be 21 V to make $V_{E3} \cong 20$ V and $V_{CC} = 40$ V, $I_{C4} = (40$ V $- 21$ V$)/10\,k\Omega = 1.9$ mA $\cong I_{E4}$. Letting $R_{E4} = 3\,k\Omega$ puts $Q4$'s emitter at 5.7 V $\simeq (1.9$ mA $\times 3\,k\Omega)$ and $Q4$'s base is at about 6.2 V $\simeq (5.7$ V $+ 0.5$V). Letting $R_{C5} = 10\,k\Omega$ then means that about $(40$ V $- 6.2$ V$)/10\,k\Omega = 3.58$ mA of dc collector current flows in $Q5$. This assumes that negligible current need be supplied to the base of $Q4$. This is a valid assumption since $I_{B4} = I_{C4}/h_{fe4} \simeq 2$ mA$/100 = 0.02$ mA, which is small in comparison to I_{RC5} (3.58 mA). Letting R_{E5} equal 500 Ω puts $Q5$'s emitter at about 3.58 mA (0.5 kΩ) $\simeq 1.8$ V, which means that $R1$ should be adjusted to provide about 2.3 V at the base of $Q5$. If we let $R2 = 5\,k\Omega$, then a ballpark value for R_A can be calculated. Since $I_{C5} \simeq 3.58$ mA dc, $I_{B5} \simeq 3.58$ mA$/100 = 0.0358$ mA. $I_{RB} \simeq 2.3$ V$/5\,k\Omega = 0.46$ mA. Therefore, R_A must supply a current of $I_{B5} + I_{R2} = 0.4958$ mA $\simeq 0.5$ mA and should drop 40 V $- 2.3$ V $= 37.7$ V. Therefore,

$$R_A = \frac{37.7\,\text{V}}{0.5\,\text{mA}} = 75\,k\Omega$$

Experimentally, R_A could be adjusted using a potentiometer while monitoring the collector voltage of $Q5$. When V_{C5} is at 21 V, the value of R_A may be measured with an ohmmeter.

Figure 9-11 shows the complete amplifier with all resistor values. All

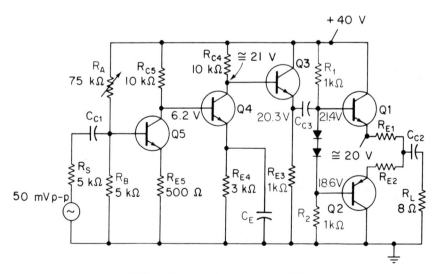

FIG. 9-11. Complete power amplifier

that remains now is to determine the appropriate capacitor sizes and to calculate G_v to ensure conformance to the initially given specifications. The voltage gain of the first stage, $Q5$, is

$$(R_{C5} \| R_{IN4})/R_{E5} = \frac{10 \text{ k}\Omega \| (h'_{fe} \times h_{ib4})}{500 \text{ }\Omega}$$

and h_{ib4} is about 50 Ω; therefore,

$$G_{v5} = \frac{10 \text{ k}\Omega \| 5 \text{ k}\Omega}{500 \text{ }\Omega} \simeq 6.7$$

The voltage gain of $Q4$ is $(R_{C4} \| R_{IN3})/h_{ib4}$, or

$$G_{v4} = \frac{10 \text{ k}\Omega \| 25 \text{ k}\Omega}{50 \text{ }\Omega} = 143$$

Assuming a gain of 1 for the class AB emitter followers yields an overall voltage gain, G_{voa}, of

$$\begin{aligned} G_{voa} &= G_{v5} \times G_{v4} \times \frac{Z_{IN5}}{Z_{IN5} + R_S} \\ &\simeq 6.7 \times 143 \times \frac{5 \text{ k}\Omega}{5 \text{ k}\Omega + 3 \text{ k}\Omega} \\ &\simeq 600 \end{aligned}$$

Thus we have a little more gain than is actually required. This could easily be compensated for by some series resistance in series with R_S or a potentiometer level control.

We should now ensure that clipping does not occur at any point in the amplifier. With 50 mV p-p input, the voltage divider between R_S and Z_{IN5} applies about 31 mV p-p to $Q5$'s base, and thus 31 mV $\times$ G_{v5} = 21 mV $\times$ 6.7 = 207 mV p-p at $Q5$'s collector. Even though $Q5$ is not biased anywhere near its mid bias point of 20 V, the ac signal at its collector is so small that no clipping will occur. $Q4$ amplifies the 207 mV signal by a factor of 143. Thus the ac signal at $Q4$'s collector is 29.6 V p-p. Since its collector is at 21 V, clipping will not occur. However, recall that only 25.3 V p-p was the requirement to provide 10 W output, and at that level the amplifier operation should be quite good.

To determine capacitor sizes, we shall let the speaker coupling capacitor, C_{C2}, determine the low-frequency break of 20 Hz. It will be the largest and most expensive capacitor because it is working in conjunction with such a low impedance:

$$f_2 = 20 \text{ Hz} = \frac{1}{2\pi R C_{C2}}$$

The resistance seen by C_{C2} is $8\,\Omega$ plus $R_{E1\text{ or }2} + R_{C4}/h'_{fe3}h'_{fe1} + 1\,\text{k}\Omega\,\|\,1\,\text{k}\Omega\,\|\,1\,\text{k}\Omega/h'_{fe\,1\text{ or }2}$

$$20\text{ Hz} = \frac{1}{2\pi[8\,\Omega + 8\,\Omega + 1\,\Omega + 333\,\Omega/100\,[10\,\text{k}\Omega/(100\times 100)]C_{C2}}$$

Therefore,

$$20\text{ Hz} \cong \frac{1}{2\pi(13\,\Omega)C_{C2}}$$

Solving for C_{C2} yields

$$C_{C2} = \frac{1}{2\pi(13\,\Omega)20}$$
$$= 610\,\mu\text{F}$$

The bypass capacitor, C_E, "sees" $3\,\text{k}\Omega$ in parallel with $h_{ib4} + (10\,\text{k}\Omega/100) = 3\,\text{k}\Omega\,\|\,(50 + 100) \simeq 150\,\Omega$. Select C_E to cause a break frequency at 5 Hz so that it will not affect the overall value of f_1. Then

$$5\text{ Hz} = \frac{1}{2\pi(150\,\Omega)C_E}$$

$$C_E = \frac{1}{2\pi \times 5 \times 150\,\Omega} = 212\,\mu\text{F}$$

The cost of C_{C2} would typically be four times that of C_E not only due to its greater capacity, but also because it requires a higher voltage rating (20 versus 6 V). Selecting C_{C1} to also provide a 5 Hz break frequency yields

$$5\text{ Hz} \cong \frac{1}{2\pi C_{C1}(5\,\text{k}\Omega + 3\,\text{k}\Omega)}$$

$$C_{C1} = \frac{1}{2\pi 5\text{ Hz}(8\,\text{k}\Omega)}$$

$$C_{C1} \cong 4\,\mu\text{F}$$

The coupling capacitor C_{C3} sees a resistance of

$$R_{E3}\,\|\,(h_{ib3} + R_{C4}/h_{fe3}) + (R_1\,\|\,R_2 h'_{fe1\text{ or }2} \times 9\,\Omega)$$
$$= 1\,\text{k}\Omega\,\|\,(50\,\Omega + 10\,\text{k}\Omega/100) + (1\,\text{k}\Omega\,\|\,1\,\text{k}\Omega\,\|\,900\,\Omega)$$
$$\cong 450\,\Omega$$

Once again we will select a capacitor to provide a 5 Hz cut off so

$$C_{C3} = \frac{1}{2\pi \times 5\text{ Hz} \times 450\,\Omega}$$
$$\cong 70\,\mu F$$

The preliminary design is now complete (on paper at least) and is ready for laboratory breadboarding and testing. The component values computed above may be modified somewhat as a result of a variety of tests involving changes in temperature, frequency response, input signal voltage, and so forth.

PROBLEMS

1. Determine G_{voa}, G_{ioa}, Z_{IN}, e_{out}, and Z_O for the amplifier system of Fig. P9-1.

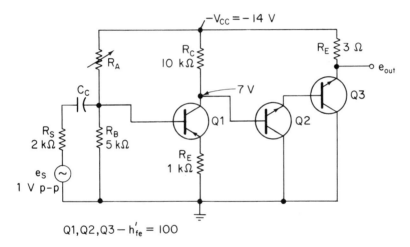

FIG. P9-1.

2. If R_{E3} in Problem 1 were changed to 1 Ω, what would be the new value of e_{out}?
3. Design a multistage amplifier that transforms a 2 mV p-p, 1 MΩ source into a 1 V p-p signal with 2 kΩ output impedance. Field-effect transistors are not allowable in this design.
4. Design the previous amplifier with an FET input stage.
5. For the amplifier of Fig. 9-6, determine the range of $Q3$ and $Q4$'s Q-point voltage if $Q3$ has an I_{DSS} range of from 8–15 mA. Is this an acceptable situation?
6. Determine G_v and G'_v for the amplifier of Fig. P9-6. Both transistors have $h'_{fe} = 100$ and 10 V Q-points.
7. Determine the input impedance and current gain of Problem 6.
8. Determine the low-frequency cutoff of the amplifier of Fig. P9-6 with and without the negative feedback connection. [Hint: Refer to Eq. (8-11)].

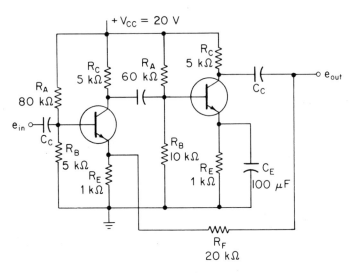

FIG. P9-6.

9. Repeat Problem 8 for the high-frequency cutoff, given that $C_{be} = 5$ pF and $C_{bc} = 8$ pF for both transistors. [Hint: Refer to Eq. (8-12)].

10. What would f_1 be for the amplifier of Fig. 9-8 if the bypass capacitor were changed to 1000 μF?

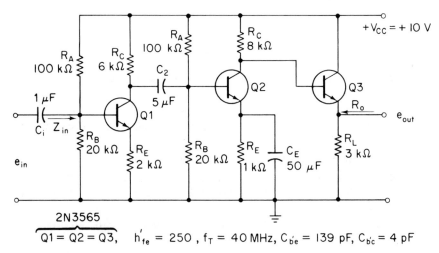

2N3565
$Q1 = Q2 = Q3$, $h'_{fe} = 250$, $f_T = 40$ MHz, $C_{b'e} = 139$ pF, $C_{b'c} = 4$ pF

$I_E = 1$ mA for all transistors

FIG. P9-11.

11. Analyze the circuit of Fig. P9-11 for G_{voa}, Z_{in}, R_o, and its 3 dB down frequencies.

12. Design a 5 W audio amplifier to amplify from a 5 mV p-p source with 10 kΩ internal impedance. A low-frequency response down to 30 Hz is required. You are to specify *all* values of all components.

10

Oscillators

10-1 Introduction

An oscillator is a circuit capable of converting energy from a dc form to an ac form. We shall concern ourselves with the generation of sinusoidal waveforms here, but nonsinusoidal waveform generation finds much practical usage also. They are most appropriately covered in a digital circuits course, however.

One basic application for an oscillator is in a communication transmitter where the audio information is combined with a high frequency signal that is generated by an oscillator. The purpose of this is to allow transmission of low frequency information at a higher carrier frequency, since transmission of audio frequencies directly is impractical for a number of reasons. Modern communication receivers of the superheterodyne variety also require an oscillator to allow a stepdown to the intermediate frequency (IF). An introduction to communications theory and equipment is provided in Chapter 13.

Another important oscillator application is in electronic test equipment. To successfully check out many circuits, it is necessary to apply an oscillator output of correct waveform, frequency, and voltage level. Test generators (oscillators) whose output level and frequency can be varied over a wide

226 Oscillators

range are commonly used for this purpose. They can often supply several types of waveforms, such as sine waves, square waves, triangular, and sawtooth waveforms.

A number of different forms of sine wave oscillators are available for use in electronic circuits. They fall into the following broad classifications and form the basis of study in this chapter:

1. *LC* feedback oscillators.
2. *RC* phase selective oscillators.
3. Negative-resistance oscillators.

The choice between these forms, as well as between individual varieties of these categories, is based upon the following requirements:

1. Output frequency required.
2. Frequency stability required.
3. Is the frequency to be variable and, if so, over what range?
4. Allowable waveform distortion.
5. Power output requirement.

These performance considerations combined with economic factors will then dictate the form of oscillator to be used in a given application.

10-2 Ringing in an IC Tank Circuit

A very close analogy exists between a swinging pendulum and a tank circuit. It is often easier for the student to visualize an electrical phenomenon by first examining its mechanical counterpart. With reference to Fig. 10-1a, moving the pendulum from its rest position to point A provides the pendulum with *potential energy*—the energy of position. Release of the pendulum from point A starts a conversion process of potential energy into the energy of

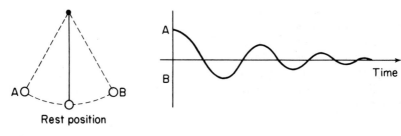

(a) Pendulum action   (b) Pendulum displacement versus time

FIG. 10-1. Pendulum: (a) pendulum action; (b) pendulum displacement versus time

motion—*kinetic energy*. The pendulum now builds up speed and then passes right on through the rest position on to point *B*, where the kinetic energy has now been reconverted into potential energy. Since losses are introduced by friction and windage, the pendulum displacement at *B* will be somewhat less than the initial displacement at *A*. If the pendulum is allowed to continue to swing back and forth, each succeeding swing will be smaller than the preceding one until the motion ceases, as shown in Fig. 10-1b. This is known as a *damped sinusoidal oscillation* and is a sinusoidal waveform. In a clock, however, the motion is not allowed to stop but is maintained by supplying a small force to the pendulum, at the right time, to compensate for the losses. The frequency of the pendulum's swing is a constant determined by its mass and length, and it regulates the speed of the clock's hands quite accurately.

The effect of charging the capacitor in Fig. 10-2a to some voltage potential and then closing the switch results in the waveform shown in Fig. 10-2b,

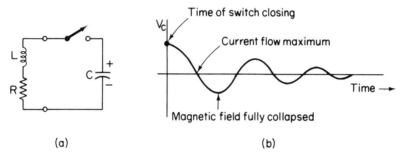

FIG. 10-2. Tank circuit "flywheel" effect: (a) tank circuit; (b) "flywheel" effect

which is identical to the one shown in Fig. 10-1b for the pendulum. The switch closure (which corresponds to releasing the pendulum) starts a current flow as the capacitor begins to discharge through the inductor. The inductor, which resists a change in current flow, causes a gradual sinusoidal current buildup that reaches a maximum when the capacitor is fully discharged. (This corresponds to the pendulum just reaching its rest position after its initial release.) At this point the potential energy is zero, but since current flow is maximum, the magnetic field energy around the inductor is maximum (this corresponds to the kinetic energy of the pendulum). The magnetic field no longer maintained by capacitor voltage then starts to collapse, and its counter emf will keep current flowing in the same direction, thus charging the capacitor to the opposite polarity of its original charge. The circuit losses (mainly the dc winding resistance of the coil) cause the output to become gradually smaller as this process repeats itself after the complete collapse of the magnetic field. The energy of the magnetic field has been converted into the energy of the capacitor's electric field and vice versa. The process repeats itself at the *natural* or *resonant* frequency f_o, as predicted by Eq. (8-15):

$$f_o = \frac{1}{2\pi\sqrt{LC}} \tag{8-15}$$

In order for an *LC* tank circuit, as shown in Fig. 10-2, to function as an oscillator, an amplifier is utilized to restore the lost energy to provide a constant amplitude sine wave output. The resulting "undamped" waveform is known as a *continuous wave* (CW) in radio work. In Section 10-3 the most straight-forward method of electronically restoring this lost energy is examined, and the general conditions required for oscillation are introduced.

10-3 Basic LC Oscillators

The *LC* oscillators are basically feedback amplifiers with the feedback serving to increase or sustain the self-generated output. This is called *positive feedback*, and it occurs when the fed back signal is in phase with (reinforces) the input signal and the term $BG_v = 1$ in the general feedback equation [Eg. (5-4)]:

$$G'_v = \frac{G_v}{1 - BG_v} \tag{5-4}$$

The closed-loop voltage gain G'_v is unstable at this point, since the denominator of Eq. (5-4) is equal to zero. It would seem then that the regenerative effects of this positive feedback would cause the output to continually increase with each cycle of fed back signal. However, in practice, component nonlinearity and power supply limitations limit the theoretically infinite gain when $BG_v = 1$. The term BG_v in Eg. (5-4) is usually made slightly larger than 1 to minimize the inherent instability that exists when BG_v equals exactly 1. This also serves to compensate for any circuit variations that might cause BG_v to become less than 1, an effect that would cause the oscillations to die out.

The criteria for oscillation that have just been outlined are formally stated by the *Barkhausen criteria* as follows:

1. The loop gain BG_v must be slightly greater than 1.
2. The loop phase shift must be $n(360°)$, where $n = 1, 2, 3, \ldots$.

An oscillating amplifier adjusts itself to meet both of these criteria. The initial surge of power to such a circuit creates a sinusoidal voltage in the tank circuit at its resonant frequency, and it is fed back to the input and amplified repeatedly until the amplifier works into the saturation and cutoff regions. At this time the flywheel effect of the tank is effective in maintaining a sinusoidal output. This process shows us that too much gain would cause excessive impurity (distortion) of the waveform and hence should be limited to a level which is just greater than 1. This is necessary to maintain oscillations under all possible conditions.

The circuit shown in Fig. 10-3 will now be analyzed to show the basics of oscillator action. Oscillators of this form are known as *Franklin* oscillators.

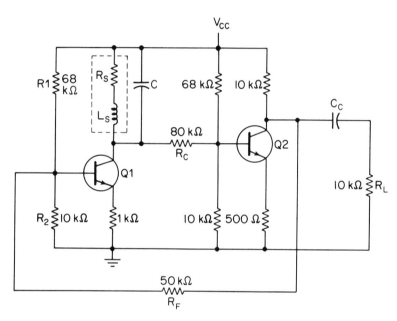

FIG. 10-3. Franklin oscillator

Note that two CE stages are utilized, which result in two 180° phase shifts or a total of 360° from input to output. Thus the phase-shift criteria for an oscillator is satisfied. Standard techniques will be used to determine whether a loop gain greater than 1 has been accomplished.

The initial application of power to this circuit will start the LC tank circuit ringing at its resonant frequency of $1/(2\pi\sqrt{LC})$. If $L = 0.1$ H and $C = 0.01$ μF, f_o equals 5000 Hz. If the series dc winding resistance R_S of the inductor L were measured as 1 kΩ, we could determine the impedance of the tank circuit at resonance Z_o from Eq. (8-16):

$$Z_o = \frac{L_S}{CR_S}$$

$$= \frac{0.1 \text{ H}}{0.01 \times 10^{-6}\text{F} \times 1 \text{ k}\Omega} = 10 \text{ k}\Omega$$

(8-16)

Thus $G_{v1} \simeq 10\,\mathrm{k\Omega}/1\,\mathrm{k\Omega} = 10$ and $G_{v2} \simeq (10\,\mathrm{k\Omega} \| 10\,\mathrm{k\Omega})/500\,\Omega = 10$, and the loop gain is 100 not including the attenuating effects of R_C, the coupling resistor between stages, and R_F, the feedback resistor. These two high-valued resistors serve to reduce the loop gain to just greater than 1, as well as to isolate the bias points of the two stages as coupling capacitors do.

The use of coupling capacitors would change the loop phase shift away from 360° and would mean that the 360° phase-shift requirement would be fulfilled at some off-resonant frequency. The resulting phase shift of the tank circuit would then compensate for the coupling capacitor phase shift, but distortion and instability problems would result.

In going from stage 1 to stage 2 the 80 kΩ coupling resistor feeds into $10\,\mathrm{k\Omega} \| 68\,\mathrm{k\Omega} \| h'_{fe2}\, 500\,\Omega$ or 7.5 kΩ. Thus the voltage attenuation is $7.5\,\mathrm{k\Omega}/(80\,\mathrm{k\Omega} + 7.5\,\mathrm{k\Omega})$ or about 0.085. The 50 kΩ feedback resistor R_F is feeding into roughly the same impedance and results in $7.5\,\mathrm{k\Omega}/(50\,\mathrm{k\Omega} + 7.5\,\mathrm{k\Omega})$ or about a 0.13 attenuation. The resulting overall loop gain is the product of all voltage gains and attenuations, or

$$G_{voa} = 10 \times 10 \times 0.085 \times 0.13 \cong 1.1$$

Thus the loop gain requirement of greater than 1 is just satisfied. However, the calculations are only approximate, and laboratory testing would be desirable to insure satisfactory performance.

10-4 Hartley, Colpitts, and Clapp Oscillators

The LC oscillators discussed in this section function on the same principles as the Franklin oscillator, but their action is somewhat more difficult to visualize. Their names are derived from some of the early pioneers in the field of radio, who developed these oscillators.

Figure 10-4 shows the basic Hartley oscillator in simplified form. The inductors $L1$ and $L2$ are a single center-tapped inductor. Positive feedback is obtained by mutual inductance effects between $L1$ and $L2$, with $L1$ in the transistor output circuit and $L2$ across the base emitter or input circuit. A portion of the amplifier signal in the collector circuit ($L1$) is returned to the base circuit by means of inductive coupling from $L1$ to $L2$. As always in a CE circuit, the collector and base voltages are 180° out of phase. Another 180° phase reversal between these two voltages occurs because they are taken from opposite ends of an inductor (the $L1$–$L2$ combination) with respect to the inductor tap that is tied to the common transistor terminal–the emitter. Thus the in-phase feedback requirement is fulfilled and loop gain is of course

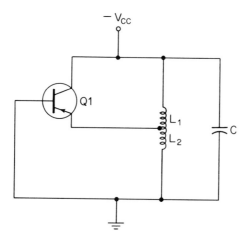

FIG. 10-4. Simplified Hartley oscillator

provided by $Q1$. The frequency of oscillation is approximately given by

$$f_o = \frac{1}{2\pi\sqrt{L1 + L2C}} \qquad (10\text{-}1)$$

f_o is influenced slightly by the transistor parameters and amount of coupling between $L1$ and $L2$.

Figure 10-5 shows a practical Hartley oscillator. A number of additional circuit elements are necessary to make a workable oscillator over the simplified one we used for explanatory purposes in Fig. 10-4. Naturally, the resistors R_A and R_B are for biasing purposes. The radio frequency choke (RFC) is effectively an open circuit to the resonant frequency and thus allows a path for the bias current, but does not allow the power supply to short out the ac signal. The coupling capacitor $C3$ prevents dc current from flowing in the tank, and $C2$ provides dc isolation between the base and the tank circuit. Both $C2$ and $C3$ can be considered as short circuits to the oscillator's frequency.

Figure 10-6 shows a Colpitts oscillator. It is similar to the Hartley oscillator except that the tank circuit elements have interchanged their roles. The capacitor is now split, so to speak, and the inductor is single valued with no tap. The details of circuit operation are identical with the Hartley oscillator and therefore will not be further explained. The frequency of operation is given approximately by the resonant frequency of the $L1$ and $C1$ in series with $C2$ tank circuit:

$$f_1 = \frac{1}{2\pi\sqrt{C1C2/(C1 + C2)L1}} \qquad (10\text{-}2)$$

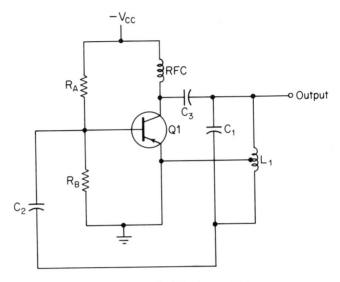

FIG. 10-5. Practical Hartley oscillator

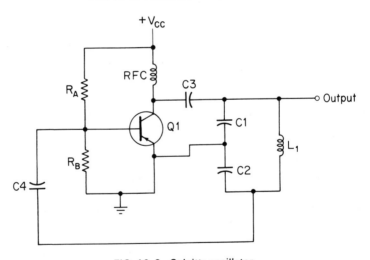

FIG. 10-6. Colpitts oscillator

The performance differences between these two oscillator forms are minor, and the choice between them is usually made on the basis of convenience or economics. They may both provide variable oscillator output frequencies by making one of the tank circuit elements variable.

A variation of the Colpitts oscillator is shown in Fig. 10-7. The Clapp oscillator has a capacitor $C3$ in series with the tank circuit inductor. If $C1$ and $C2$ are made large enough, they will "swamp" out the transistor's inherent

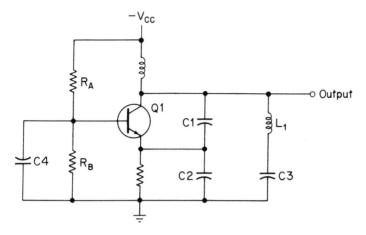

FIG. 10-7. Clapp oscillator

junction capacitances, thereby negating transistor variations and junction capacitance changes with temperature. The signal frequency is determined by the series resonant frequency of $L1$ and $C3$ when the series capacitance combination of $C1$ and $C2 \gg C3$. Thus the resonant frequency is

$$f_1 \simeq \frac{1}{2\pi\sqrt{L1C3}} \qquad (10\text{-}3)$$

and an oscillator with better frequency stability than the Hartley or Colpitts versions results. The possible range of frequency adjustment is not as large, however, with the Clapp oscillator.

The three LC oscillators presented in this section are the ones most commonly used. However, many different forms and variations exist and are used for special applications.

10-5 Crystal Oscillators

When greater frequency stability then that provided by LC oscillators is required, a crystal-controlled oscillator is often utilized. A *crystal oscillator* is one that uses a piezoelectric crystal as the inductive element of an LC circuit. The crystal, usually quartz, also has a resonant frequency of its own, but optimum performance is obtained when it is coupled with an external capacitance. Applying an alternating electric potential across the two faces of the crystal results in mechanical vibrations that have maximum amplitude at the natural resonant frequency of the crystal.

The electrical equivalent circuit of a crystal is shown in Fig. 10-8. It represents the crystal by a series resonant circuit (with resistive losses) in parallel with a capacitance C_P. The resonant frequencies of these two resonant circuits (series and parallel) are quite close together (within 1 per cent), and hence the impedance of the crystal varies sharply within a narrow frequency range.

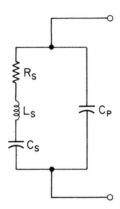

FIG. 10-8. Electrical equivalent circuit of a crystal

This is equivalent to a very high Q circuit, and in fact crystals with a Q factor of 20,000 are common; a Q of up to 10^6 is possible. This compares with a maximum Q of about 1000 with high quality inductors and capacitors. For this reason, and because of the good time and temperature stability characteristics of quartz, crystals are capable of maintaining a frequency to ± 0.001 per cent over a fairly wide temperature range. The ± 0.001 per cent term is equivalent to saying ± 10 parts per million (ppm), and this is a preferred way of expressing such very small percentages. Over very narrow temperature ranges or by maintaining the crystal in a small temperature-controlled oven, stabilities of ± 1 ppm are possible. Crystals are fabricated by "cutting" the crude quartz in a very exacting fashion. The method of "cut" is a science in itself and determines the crystal's natural resonant frequency as well as its temperature characteristics. Crystals are available at frequencies of about 15 kHz and up, with the higher frequencies providing the best frequency stability. However, at frequencies above 10 MHz they become so physically small that handling becomes a problem.

Crystals may be used in place of the inductors in any of the previously discussed *LC* oscillators. A circuit especially adapted for crystal oscillators is the Pierce oscillator shown in Fig. 10-9. The use of an FET is desirable, because the light loading of the crystal provides for good stability and does not lower the Q. This circuit is essentially a Colpitts oscillator with the crystal replacing the inductor and the inherent FET junction capacitances

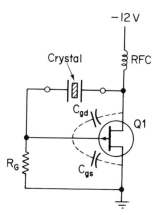

FIG. 10-9. Pierce oscillator

functioning as the split capacitor. Because these junction capacities are generally low, this oscillator is effective only at high frequencies.

A special advantage of the Pierce oscillator is the simple means of changing frequency. Since there are no tuned circuits, as such, the frequency can be changed by plugging a different crystal into the circuit at points X and Y in Fig. 10-9. This is a useful feature for many radio transmitters that are required to work at a number of different specific frequencies.

10-6 RC Phase-Shift Oscillators

If a sinusoidal oscillator is necessary for frequencies below about 10 kHz, LC oscillators become impractical. This is due to the bulk and expense of the high-valued inductors required. An alternative to the creation of sine waves via tank circuits is the use of RC selective filters. The RC phase-shift oscillator of this section and the Wien bridge oscillator of the next are the two most common examples of RC selective filters.

Initially, one may think that the formation of a sine wave is impossible with the circuit of Fig. 10-10. The ability of this circuit to oscillate sinusoidally is based upon the fact that only one frequency can pass through the RC phase-shifting network with 180° of phase shift. This phase shift coupled with the transistor's 180° input–output phase shift means that only one sinusoidal frequency can successfully fulfill the Barkhausen phase shift requirements, and hence regenerative effects occur for one frequency only and a sinusoidal output is possible.

Since the phase-shifting network must supply 180° phase shift, at least

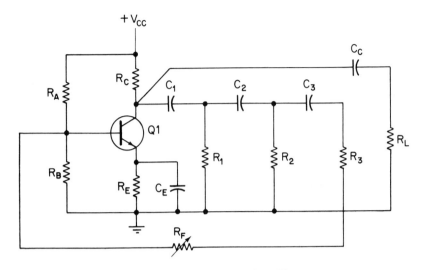

FIG. 10-10. *RC* phase-shift oscillator

three *RC* sections must be used because a maximum of only 90° phase shift can be approached per section. Often four are used, since the attenuation introduced is actually less with four than with three sections. The resistance of the last *RC* network (R_3–C_3 in the case of Fig. 10-10) also has the input impedance of the bias resistors and $Q1$ as a load in determining its phase shift.

The attenuation of the *RC* circuits must be compensated for by the gain of the transistor to allow a total loop gain greater than 1. Unfortunately, a gain much greater than 1 results in poor stability for this circuit, and it is often necessary to adjust the circuit's gain to the proper level to obtain satisfactory results. This can be accomplished by inserting a variable resistor in the feedback path, as shown in Fig. 10-10. In any event, a fairly high h_{fe} transistor must be used to overcome the *RC* network's losses.

The frequency of oscillation for the three-section *RC* oscillator when the *R* and *C* components are equal is roughly approximated as

$$f \simeq \frac{1}{18RC} \qquad (10\text{-}4)$$

If a four-section *RC* circuit is used, the frequency of oscillation is

$$f \simeq \frac{1}{9RC} \qquad (10\text{-}5)$$

The circuit will operate with the capacitors in shunt, as shown in the four-section circuit of Fig. 10-11. The frequency of oscillation for the four-section

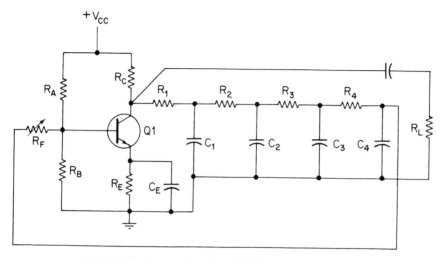

FIG. 10-11. Four-section shunt C RC phase-shift oscillator

shunt C circuit is

$$f \simeq \frac{1}{6RC} \qquad (10\text{-}6)$$

and for the three-section shunt C circuit

$$f \simeq \frac{1}{3RC} \qquad (10\text{-}7)$$

These circuits do not lend themselves to frequency adjustment over a wide range because of the large number of resistors or capacitors that would have to be varied. In addition, a gain adjustment would be necessary because of the different attenuation that such adjustments would cause. Without gain adjustment the loop gain would subsequently drop below 1, causing oscillation to cease or become so much greater than 1 so as to cause instability. Distortion levels of 5 per cent in the output signal are typical for this circuit.

10-7 Wien Bridge Oscillator

Whenever a wide range of frequencies is to be generated and a low distortion level is required, the Wien bridge oscillator shown in Fig. 10-12 is most often used. It is the circuit found in most laboratory sine-wave generators, which have a frequency range of 5 Hz–500 kHz, typically. The two capacitors in Fig. 10-12 are generally varied as a fine frequency adjust and a range

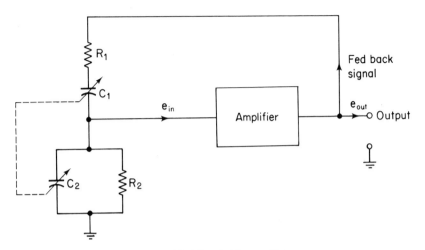

FIG. 10-12. Wien bridge oscillator

switch may be used to switch the value of all components in the series *RC* and parallel *RC* circuits to allow large frequency changes. The resistors and capacitors are usually equal in value. The dashed lines between *C*1 and *C*2 indicate that these variable capacitors are "ganged" together; i.e., they are mechanically attached such that one control knob adjusts them both simultaneously.

The basis of operation in this circuit is the fact that there is only one frequency that causes the fed back signal to have zero phase shift, and it just happens to be the frequency that has the highest amplitude fed into the amplifier's input. For very low frequencies fed back into the series *R*1–*C*1 circuit, *C*1 appears as an open circuit, thus making the voltage e_{in} very small. As the fed back signal's frequency is gradually increased, e_{in} will increase until the shunt capacitor *C*2 starts appearing as a short circuit, causing e_{in} to decrease. Thus we see that the voltage e_{in} goes through a peak value, and this frequency happens to be the only one with zero phase shift from e_{out} to e_{in}. Thus if the amplifier is noninverting, as with two CE stages cascaded together, the Barkhausen criteria are fulfilled for one frequency only and a stable oscillation results.

A mathematical solution of the series–parallel *RC* circuit shows that for equal values of *R* and *C*, the frequency of maximum fed back signal and zero phase shift is

$$f_o = \frac{1}{2\pi RC} \qquad (10\text{-}8)$$

A practical form of the Wien bridge oscillator is shown in Fig. 10-13. Notice here that the *RC* circuits are combined with two other resistors (*R*3 and *R*4)

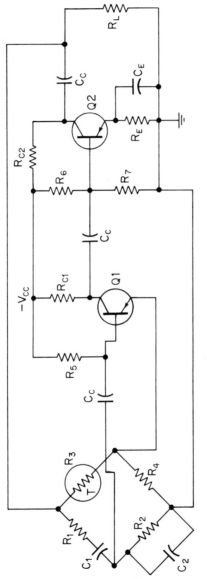

FIG. 10-13. Wien bridge oscillator

to form a bridge circuit. This provides greater sensitivity, as well as a means of preventing excessive loop gain and the resulting output instability. Consider a case when the loop gain is excessive. This causes the ac fed back signal to increase and will therefore increase the power dissipation in the thermistor, $R3$. It will heat up, lowering its resistance, thereby increasing the emitter degeneration (negative feedback) in $Q1$'s emitter resistor $R4$. The overall loop gain is therefore reduced, providing adequate circuit performance even as the frequency is adjusted and other loop gain variations occur.

You may be saying to yourself at this time that it is understandable how only one possible sinusoidal frequency can effectively oscillate in these *RC* oscillators, *but* where does *that* particular signal originate in the first place? In the initial application of dc power to the circuit a sudden voltage step is applied to the entire circuit. This step function can be thought of as being made up of a large number of sinusoidal frequencies—all these die out except for that one specific frequency that satisfies the Barkhausen criteria. It is, however, sometimes a problem to get these circuits to start oscillating, and therefore a means for greater loop gains during turn on is sometimes employed in Wien bridges.

10-8 Negative-Resistance Oscillators

Recall that a sudden application of voltage to an *LC* tank circuit causes a sinusoidal output at the tank's resonant frequency. Unfortunately, the exchange of energy between the inductor and capacitor that generates the sine wave gradually dies out, due to dissipation in the coil's winding resistance. If there were some way of canceling this resistance, it would seem likely that a steady oscillation could be maintained; this is the principle of sinusoidal negative-resistance oscillators. The energy to maintain the oscillations is supplied to the circuit by the dc power source with the negative-resistance element allowing this transfer of energy to the tank circuit.

The two generally used negative-resistance semiconductor devices are the tunnel diode and the unijunction transistor. They are both useful in nonsinusoidal applications in circuits known as relaxation oscillators. These are most appropriately covered in a switching circuits course. The tunnel diode is also used to generate sine waves—usually at very high frequencies. In fact, operation at above 100 GHz (10^{11} Hz) has been reported with tunnel diodes.

The tunnel diode is a semiconductor *pn* junction that is very heavily doped, perhaps 1000 times more than in ordinary diodes. This results in an extremely thin depletion layer that exhibits a *tunneling effect*, resulting in a negative resistance behavior. This effect was developed by L. Esaki in 1957. Figure 10-14 shows some common symbols for the tunnel diode, its equiva-

Negative-Resistance Oscillators

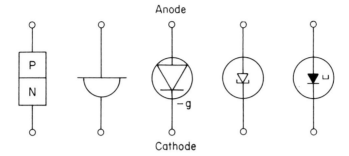

(a) Tunnel diode symbols

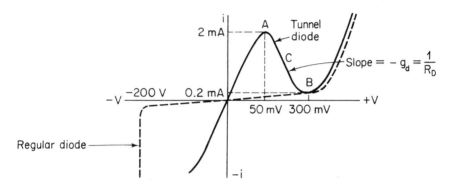

(b) Germanium diode characteristics and tunnel diode characteristics

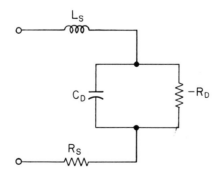

(c) Tunnel diode equivalent circuit

FIG. 10-14. Tunnel diode

lent circuit, and its characteristic curve. The tunnel diode exhibits negative resistance when biased in the region A–C–B between A and B in Fig. 10-14b. Figure 10-14c shows the equivalent circuit when so biased with $-R_D$ being the negative resistance. The resistance presented by the diode in this region is negative, because an increase in diode voltage results in a decrease in current flow. Typical values for the elements of the equivalent circuit are $L_S = 0.1$ nH, $C_D = 2.5$ pF, $R_S = 1\ \Omega$, and $R_D = -50\ \Omega$.

Figure 10-15 shows a simple tunnel diode oscillator circuit. If properly biased and if the diode's negative resistance is greater than R_S, oscillation will begin when power is applied, with the change in diode current oscillating between points A and B on Fig. 10-15b. As the current amplitude swing

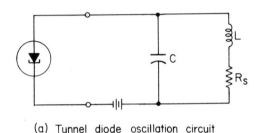

(a) Tunnel diode oscillation circuit

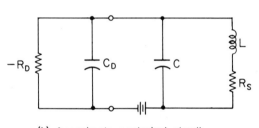

(b) Appoximate equivalent circuit

FIG. 10-15. Tunnel diode oscillator

gets bigger, the diode's operating region starts extending past points A and C where the negative resistance starts decreasing. This causes signal stabilization at the amplitude limits, where $|R_D| = R_S$, and the net circuit resistance is negligible.

The frequency of oscillation for these circuits is dependent upon the tank circuit elements and the diode capacitance, C_D, and is

$$f_o = \sqrt{\frac{1}{L(C + C_D)} - \frac{1}{R_D^2 C_D(C_D + C)}} \qquad (10\text{-}9)$$

Unfortunately, the diode capacitance is highly dependent on the bias voltage and temperature, and these oscillators tend to be unstable. However, if a very high Q tank is *loosely coupled* to the diode, a reasonably stable oscillator, independent of diode parameter variations, results.

10-9 Parasitic Oscillations

Unfortunately, the Barkhausen criteria are sometimes fulfilled when it is not desirable to do so. This leads to oscillations that are referred to as *parasitic* or unwanted. They can occur in any circuit that has gain and typically are of a very high frequency. This is because the capacitances and/or inductances causing these effects are usually low in value. Connecting leads have a small inductance that can present a high inductive reactance at high frequencies or can resonate with small associated stray wiring capacitances at high frequencies. If the layout of these reactances is such as to cause some positive feedback in a high gain amplifier, parasitic oscillations will result.

These unwanted oscillations will cause distortion of the desired signal and will often render a circuit useless. Parasitic oscillations are usually the most troublesome in early stages of an amplifier, since they are then subjected to a high gain and become quite large in the output. They may also become radiated into space and interfere with other nearby electronic equipment. To prevent these effects, it is necessary, first, to lay out high gain circuits with care so as to minimize all lead lengths and to carefully watch the spacings between leads and components that may develop capacitive coupling. This is a difficult process, and sometimes even extreme care does not eliminate the parasitics. In these cases it is often helpful to isolate the approximate location of the problem and to then connect a low-valued capacitor (0.001 μF or less) between various points in this circuit. This has the effect of either neutralizing the high frequency positive feedback or reducing the high frequency gain of the amplifier to the point where a loop gain of less than 1 exists for the parasitic oscillation. This, of course, may be an unacceptable solution if the amplifier is required to have a frequency response in the same vicinity that the parasitic exists in.

Low frequency parasitics are also sometimes a problem in multistage amplifiers being powered from a single dc power source. The source of these unwanted oscillations is a mutual coupling effect, through the power supply, between the various stages. The common solution to this problem is to introduce decoupling techniques between the various offending stages. This takes the form of placing a filter capacitor across the offending stage at its physical location and electrically in parallel with the power supply's output. Even greater decoupling will result if a low-valued resistor is placed in series between the power supply and the shunt capacitor. This has the effect of lowering the voltage regulation to that stage, however. Figure 10-16 illustrates this technique with $R1$, $R2$, $C1$, and $C2$ providing the decoupling.

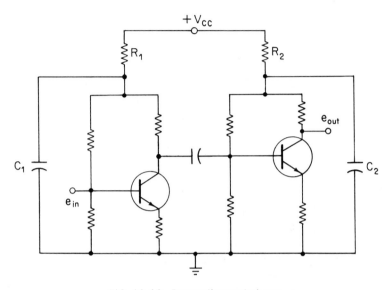

FIG. 10-16. Decoupling techniques

One other cause of unwanted low-frequency oscillations is the grounding of various points in an amplifier at many different points on a metal chassis. This results in circulating ground currents. Although the resistance of the chassis metal is quite small, there is still a slight IR drop between separate ground points. This then allows the power supply ripple voltage to be coupled into an input stage, with it then appearing amplified in the output. To solve this problem, it is necessary to use a common ground point, i.e., to connect all circuit grounds to a common point on the chassis.

PROBLEMS

1. Design a Franklin oscillator to provide an output at 100 kHz. You are to use an inductor with $L = 5$ mH and a dc winding resistance of 20 kΩ.
2. Calculate f_o for a Colpitt's oscillator using 0.1 μF capacitors and a 10 mH choke.
3. Express a ± 0.1 ppm tolerance as a percentage.
4. Calculate f_o for a three-section shunt R RC phase-shift oscillator when $R = 1$ kΩ and $C = 0.5$ μF. Repeat for a four-section oscillator.

5. Repeat Problem 4 for a shunt C RC oscillator.

6. Prove that the attenuation of a four-section RC oscillator is less than a three-section version.

7. Draw a graph of e_{out}/e_{in} versus frequency for the Wien bridge circuit of Fig. P10-7. Also draw a graph of the input-to-output phase shift versus frequency.

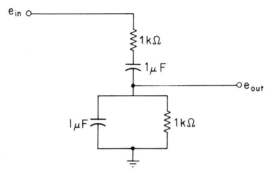

FIG. P10-7.

8. Calculate f_o for a tunnel diode oscillator working into a 50 pF, 50 μH tank. The tunnel diode has junction capacitance of 20 pF and $R_D = 50\,\Omega$.

11

Linear Integrated Circuits

11-1 Introduction

Integrated circuits (ICs) have had a profound effect on all phases of electronics in the past few years. Integrated circuits became commercially available in the early 1960s and they quickly snowballed into a giant industry in a few short years. A *monolithic* integrated circuit is a device in which a complete circuit (including all of its components) is formed upon or within a single piece of silicon crystalline material. A *thick-film hybrid* integrated circuit has the resistors, capacitors, and wire paths "screened" onto a ceramic substrate in paste form through a mesh mask. A high-temperature baking or curing cycle follows this process, and then the other components are externally added and interconnected by wire bonds. A *thin-film hybrid* integrated circuit has films deposited either by sputtering, evaporation, or chemical vapor deposition through a mask. Conductors, resistors, and capacitors are so deposited, and all other circuit elements must be added to the thin-film circuit as with the thick-film circuits. The essential difference between thick- and thin-film circuits is not their relative thickness but rather the *method* of depositing the films.

Integrated circuits offer a tremendous reduction in size over the standard printed circuits, which use standard discrete components. Since ICs lend themselves to high-volume mass-production techniques, they also offer significant cost advantages over discrete circuits. This is especially true of monolithic ICs, with the hybrid ICs generally falling somewhere between the monolithic and discrete circuits in both size and cost considerations. In fact, hybrid circuits are, as their name implies, a blend of discrete and integrated techniques.

Thus far we have talked of integrated circuits only from the method of construction standpoint. They are also classified according to type of circuit—digital or linear. This book is concerned with linear circuits, and so is this chapter on ICs. However, it is important to realize that digital ICs represent perhaps 80 per cent of the IC dollar market, with the great majority of these circuits being utilized in the computer industry. Digital ICs lend themselves to monolithic integration, because a computer uses large numbers of identical circuits, and monolithic integrated circuits become increasingly economical as the volume demand for one specific circuit increases. Linear applications are smaller in volume and tend to require differences in performance from one system to another. Despite this difficulty, linear integrated circuits (LICs) are quickly displacing their discrete-circuit counterparts in many applications as their cost becomes competitive. They also demonstrate greater reliability, because so many external connections, a major source of circuit failure, are eliminated in ICs. Linear integrated circuits find wide application in military and industrial applications as well as in consumer products. They are used in many ways. The following list includes the bulk of these functions:

1. Operational amplifiers.
2. Small-signal amplifiers.
3. Power amplifiers.
4. Sense amplifiers.
5. RF and IF amplifiers.
6. Microwave amplifiers.
7. Multipliers.
8. Comparators.
9. Voltage regulators.

Operational amplifiers are by far the most versatile form for LICs since they can be manipulated to perform all the above functions, except for the power and microwave amplifiers. Thus they make up the bulk of LICs, and we shall study them in detail in Chapter 12.

By now the variety of electronic circuits discussed has probably left you in a state of confusion. Figure 11-1 charts all the possibilities for circuit construction. Notice the new terms MSI and LSI following the monolithic circuits. They stand for medium-scale integration (MSI) and large-scale integration (LSI), respectively. They are really extensions of the monolithic

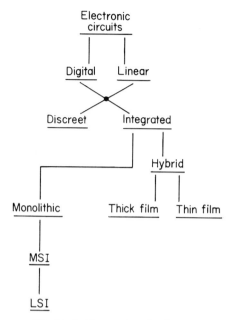

FIG. 11-1. Electronics circuit types

Table 11-1 Electronic-Circuit-Construction Cost Analysis

CLASSIFICATION		TYPICAL TOOLING COST ($)	TYPICAL UNIT COST FOR ANY SINGLE TYPE OF CIRCUIT ($)					
			1 TO 25	26 TO 100	101 TO 1000	1001 TO 10,000	10,001 TO 100,000	100,001 AND UP
Thick-film hybrid integrated circuits	digital	1,500	50	40	30	25	15	12
	linear	1,500	50	40	30	25	15	12
Thin-film hybrid integrated circuits	digital	2,000	75	60	45	30	20	15
	linear	2,000	75	60	45	30	20	15
Monolithic integrated circuits	digital	20,000	30	20	15	8	8	4
	linear	40,000	40	30	20	15	10	5
Conventional printed circuit board	digital	1,200	50	40	30	25	18	15
	linear	1,200	50	40	30	25	18	15

technology whereby whole electronic systems rather than just a circuit are incorporated in one package. They are most suited to digital circuits, and the dividing line between MSI and LSI is somewhat arbitrary. However, a good approximate rule of thumb is that MSI becomes LSI when over 100 separate circuits are incorporated and connected together in one package!

Table 11-1 provides a cost comparison for the various methods of circuit construction. It is seen that the typical tooling cost for monolithic circuits far exceeds that of any other circuit form. Thus in order for these circuits to provide a cost advantage it is necessary to be able to write off these costs with a high production volume.

EXAMPLE 11-1

Determine the total per circuit cost of a monolithic LIC, thick-film IC, thin-film IC, and conventional printed circuits for a production run of 20,000. Use the information provided in Table 11-1.

Solution:

(a) Monolithic: $\dfrac{\$40{,}000 + \$10 \times 20{,}000}{20{,}000 \text{ circuits}} = \$12$ per circuit

(b) Thick film: $\dfrac{\$1500 + \$15 \times 20{,}000}{20{,}000} = \$15$ per circuit

(c) Thin film: $\dfrac{\$2000 + \$20 \times 20{,}000}{20{,}000} = \$20$ per circuit

(d) Printed circuit: $\dfrac{\$1200 + \$18 \times 20{,}000}{20{,}000} = \$18$ per circuit

Clearly, the monolithic approach is the correct one for this high-volume application. If the requirement were for 2000 circuits, this would no longer be the case. It must be remembered that other factors may influence the decision as to the type of circuit construction other than just the economic ones. The monolithic approach offers the most compact circuit and highest reliability, with the hybrids and printed circuits in second and third place, respectively, on those accounts. On the other hand, monolithic circuits require a long time lapse from design to first production due to tooling time—typically 3 months as compared to a couple of weeks for the other forms. In addition, if changes to a circuit are required after production has commenced, as is often the case, it is virtually impossible with monolithic circuits, but this can be accomplished with the other forms. Circuits requiring precise resistor values or operation at microwave frequencies may be produced with thick-film hybrid circuits and printed circuits, but not with thin-film or monolithic circuits.

The aforementioned circuit characteristics are the major ones. The selection of circuit construction is certainly influenced by a number of other more subtle factors too numerous to pursue here. It is hoped that this discussion has at least given the student an appreciation for the complexities involved.

11-2 Monolithic IC Fabrication

The remainder of this chapter will be mainly concerned with monolithic integrated circuits. Even though industry makes widespread use of hybrid circuits for complete systems and subsystems, it is the commercial availability of low-cost monolithic LICs that is currently having the most widespread effect in today's circuits. A photograph of the typical configurations of these devices is shown in Fig. 11-2. It is shown at approximately two times actual size. The manufacturer offers these devices in a number of circuit

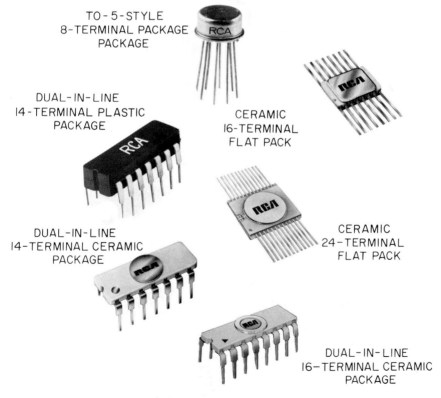

FIG. 11-2. Typical LIC configurations (2×) (Courtesy of Radio Corporation of America.)

configurations that make their use on standard printed circuit boards a common practice. Thus these standard circuit configurations (in LIC form) in conjuction with any additionally required discrete components provide the circuit designer the high degree of flexibility required for most linear circuits. It also relieves the designer from the tedious design requirements of large numbers of single-stage amplifiers, and permits him to instead design whole systems by piecing together the appropriate LIC building blocks. This is a much more efficient design approach than the 100 per cent discreet circuit and, since it will be the most often encountered form of linear circuitry for years to come, it warrants this chapter of study.

In order to skillfully deal with this monolithic–discrete component circuitry, it is helpful to first have some knowledge of how these LICs are fabricated and to learn of their special characteristics. The technology presently used for monolithic integrated circuits is based on the silicon planor techniques developed for discrete transistors. The starting point for this process is a uniform single crystal of silicon—usually p-type material, as shown in Fig. 11-3a. In this explanation a simple transistor, resistor, and capacitor circuit will be formed—those being the three possible circuit elements, but with a diode also possible by simply using either the base–emitter or base–collector junction of a transistor.

The silicon crystal has impurities introduced by a diffusion process. *Diffusion* is the introduction of controlled small quantities of material into the crystal structure. It modifies the electrical characteristics of the crystal in a tightly controlled high-temperature environment. Figure 11-3b shows the result of the diffusion of two n-type regions into the p-type substrate. The diffusion process introduces these regions to the desired depth and widths, with depth being controlled by the diffusion temperature and time, and the lateral dimensions controlled by silicon dioxide and photochemical masks. Diffusion of additional p-type and n-type regions then forms a transistor (npn) on the left; the element on the right omits the final n-type emitter diffusion, which then results in a p-type silicon resistor of controlled size (hence controlled resistance). The silicon wafer is then coated with an insulating oxide layer that is opened selectively to allow for interconnections, as shown in Fig. 11-3c. The heavy dark lines show the conductors that are formed by a metallization process. Metal (usually aluminum) is evaporated over the otherwise completed circuit and allowed to form a thin coating over the entire circuit. Then by a photosensitizing, masking, and etching process, the metal is selectively removed to leave the desired interconnect pattern.

Figure 11-3d shows a transistor–capacitor combination. A capacitor may be formed by using the oxide layer as the capacitor's dielectric with the metallization serving as the two plates. Figure 11-3e shows a transistor–resistor–capacitor combination and the schematic diagram for this circuit.

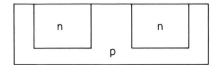

(a) Silicon wafer used as starting material for an integrated circuit.

(b) Diffusion of n-type areas to provide isolated circuit nodes.

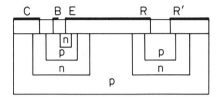
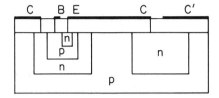

(c) Connection of contacts to p-type region to form integrated resistor.

(d) Use of oxide as a dielectric to form integrated capacitor.

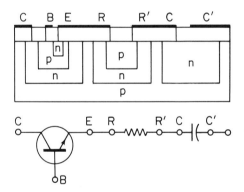

(e) Completed silicon chip containing transistor, resistor, and capacitor.

FIG. 11-3. Monolithic IC fabrication process (Courtesy of Radio Corporation of America.)

11-3 General LIC Considerations

The major cost of a monolithic IC is the tightly controlled processing steps it is subjected to. Cost is thus directly proportional to size, and the smaller the circuit the lower will be the cost. This is true because the required processing ovens can only handle a limited silicon surface area at one time and because all such ICs go through the same processes, regardless of the mix of transistors, capacitors, and resistors. Since resistors and capacitors occupy considerably more area than transistors or diodes (typically by a two or three to one factor), their use is minimized. The ratio of active to nonactive devices is therefore as high as possible in direct contrast with standard discrete circuits in which the active devices are generally the most expensive components.

Another factor that changes the design rules for IC circuits, as compared to discrete circuits, is the inability to obtain close resistor value tolerances in monolithic circuits. The resistors are of silicon material and are hence highly temperature sensitive. They do, however, drift in a predictable fashion, which thus allows for very tight resistor ratio tolerances. This characteristic can often be used to compensate for the high resistance value drift with temperature. In addition, high-valued resistances (over about 10 kΩ) are expensive from the standpoint of requiring a lot of space. This is even more critical for even moderate-sized capacitors with special capacitance multipliers used to generate capacitances of about 100 pF or greater.

Transistors in LICs have somewhat reduced high-frequency response, even though the same basic construction techniques are utilized for monolithic as will as discrete units. In addition, the ability to manufacture both *pnp* and *npn* units in the same IC is very difficult and not often done in practice. The *pnp* transistors are generally inferior in quality, even if they are the only polarity used, and we thus see a preponderance of *npn* units in monolithic circuits. However, since monolithic transistors are manufactured in very close proximity to each other and in the same processes, they generally offer very similar characteristics, and this fact is often taken advantage of.

Because of the aforementioned variations from discrete components, the monolithic circuit schematic usually takes on a strange appearance when compared to the schematic for a discrete circuit performing the same function. The inability to provide inductors in monolithic circuits is another factor that must be contended with. In summary, then, the following circuit design guidelines are applied to the design of monolithic circuits:

1. Maximize the numbers of active devices.
2. High-valued resistors and capacitors are impractical.
3. Use resistor ratios rather than absolute values.
4. Take advantage of the closely matched transistor characteristics.
5. Inductors are not available.
6. Designs incorporating only *npn* transistors are desirable.

Whenever a high-valued resistor is absolutely necessary, "pinch resistors" are utilized. They are fabricated using the same material as a transistor's base for the resistor, but by reducing its effective cross-sectional area by diffusing an emitter on top of it. The resistance is increased by not only reducing the cross-sectional area, but also by an increase in the material's resistivity due to a decreasing concentration of dopant away from the surface. The quality of these resistors is low, but they are satisfactory in uncritical applications.

11-4 Special Design Considerations

In this section some of the circuit peculiarities made possible and/or necessary by monolithic circuits are discussed.

A. DC Level Shifting

Recall that isolation between the various stages of discrete amplifiers is most often done with a coupling capacitor. These capacitors are too large to obtain in LICs, and therefore a method of dc level shifting is often employed. A level shift is possible with resistive dividers, but this of course causes undesirable signal attenuation. The circuit of Fig. 11-4 solves this problem by taking the input signal at some dc level and providing the output signal at

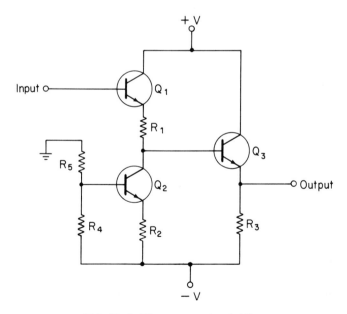

FIG. 11-4. Direct current level shifter

0V dc. This system requires dual voltage supplies, which are often available in circuits using LICs anyway.

Transistor $Q2$ in Fig. 11-4 acts as a dc constant current source since no signal is applied to its base. The value of its constant current is determined by the base bias resistors $R4$ and $R5$. Its constant current is used to provide a constant dc voltage drop across $R1$. $R1$'s dc voltage drop added to the 0.5 V dc drops across $Q1$ and $Q3$'s base–emitter junction constitutes the amount of dc level shift in the circuit.

EXAMPLE 11-2

The input signal to $Q1$ in Fig. 11-4 is 100 mV ac riding on a 10 V dc level. Calculate the ac output signal if it is riding on a 0 V dc level in order to properly drive the next stage. You are to determine all resistor values assuming the transistor's $h'_{fe} = 250$ and ± 15 V supplies are to be used.

Solution:

The required dc drop across $R1$ should be the dc level at $Q1$'s base (10 V), minus two 0.5 V base–emitter drops, minus the required dc output voltage (0 V). Hence

$$V_{R1} = 10 \text{ V} - 2(0.5 \text{ V}) - 0 \text{ V} = 9 \text{ V}$$

We must now select some resistor values. If we let $R1 = 900\ \Omega$, then the 9 V drop will mean 10 mA of dc current will flow through it. If $Q3$'s emitter is at 0 V, then $R3$ will have 15 V across it. Letting $R3 = 1.5\ \text{k}\Omega$ means the base of $Q3$ will draw 10 mA/h'_{fe3} or

$$I_{B3} = \frac{I_{C3}}{h_{fe3}} \cong \frac{10 \text{ mA}}{250} = 0.04 \text{ mA}$$

This is negligible with respect to R1's current, and therefore the constant current transistor $Q2$ must supply about 10 mA. If we let $R2 = 1\ \text{k}\Omega$, then

$$V_{R2} = I_{E3} \times R_2$$
$$\simeq 10 \text{ mA} \times 1\ \text{k}\Omega = 10 \text{ V}$$

That means V_{R5} will equal 10 V + 0.5 V, or 10.5 V, and therefore V_{R4} equals 15 V − 10.5 V = 4.5 V. Letting $R_5 = 1.05\ \text{k}\Omega$ means that R_4 must equal $0.45\ \text{k}\Omega$ by the voltage divider rule, if we can assume I_{B2} to be negligible, as is the case:

$$I_{B2} = I_{C2}/h_{fe2} \cong 0.04 \text{ mA}/250 = 0.00016 \text{ mA}$$

Figure 11-5 summarizes the selected resistor values and the resulting dc levels. Now to determine the output signal we shall first assume the voltage gains through $Q1$ and $Q3$ (both emitter followers) to be 1. The only source of signal loss in the circuit is then through $R1$. It attenuates the signal reaching $Q3$'s base by voltage divider action between $R1$ and the parallel impedance of looking into $Q3$'s base and

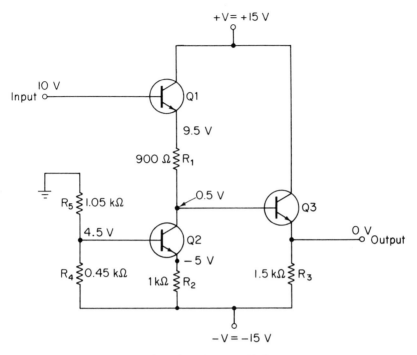

FIG. 11-5. Example 11-2

$Q2$'s collector. These are both very high impedances of over 100 kΩ, and hence the output signal equals the input signal for all practical purposes.

B. Multiplication of Capacitance

Whenever a larger capacitor is needed than is possible to fabricate in a LIC, a capacitance multiplier may be utilized. Recall from our study of the Miller effect that the effective value of a capacitor C connected between the input and output of a phase-inverting amplifier is multiplied by 1 plus the amplifier's voltage gain (Section 8-4). Hence the circuit of Fig. 11-6 could be used to accomplish this goal. Notice that the voltage amplifier, $Q2$, is buffered from the input by $Q1$ to keep from having the resulting capacitance shunted by a low resistance. Thus the resistance seen looking into $Q1$'s base is $h'_{fe1}(h_{ib1} + h'_{fe2}h_{ib2})$, and this should be high enough to provide a fairly high quality capacitance.

EXAMPLE 11-3

Calculate the input resistance and effective input capacitance for the circuit of Fig. 11-6 if $h'_{fe1} = h'_{fe2} = 250$, $R1 = 10$ kΩ, $C = 100$ pF, and $I_E = 1$ mA.

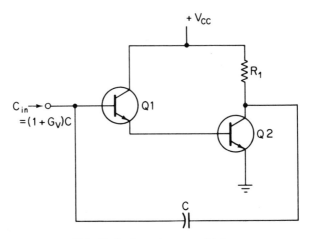

FIG. 11-6. Capacitance multiplier

Solution:
The voltage gain of this circuit is

$$G_{voa} = G_{v1} \times G_{v2} \simeq 1 \times \frac{10 \text{ k}\Omega}{h_{ib2}}$$

$$\simeq \frac{10 \text{ k}\Omega}{0.026/1 \text{ mA}} = \frac{10 \text{ k}\Omega}{26 \Omega} \simeq 400$$

Therefore, C_{in} is

$$C_{in} = (1 + G_{voa})C$$
$$\simeq 400 \times 100 \text{ pF}$$
$$= 4000 \text{ pF} = 0.04 \text{ } \mu\text{F}$$

The resistance seen in shunt with the capacitor is

$$R_{IN} = h'_{fe1}(h_{ib1} + h'_{fe2}h_{ib2})$$

and h_{ib2} is 0.026/1 mA or 26 Ω. The value of I_{E1} is 1 mA/$h_{fe2} \simeq$ 0.04 mA. This means h_{ib1} = 0.026/0.04 mA = 650 Ω. Therefore,

$$R_{IN} = 250[26 \text{ } \Omega + 250(650 \text{ } \Omega)]$$
$$\simeq 40 \text{ M}\Omega$$

C. Elimination of Emitter Bypass Capacitor

Common-emitter (CE) amplifiers offer high gain if no emitter resistance is used. Recall that for stability reasons an emitter resistor is desirable, but is used at the expense of gain, however. The use of an ac short (a capacitor) across the emitter resistor restores that gain potential, but unfortunately a large capacitor is needed if good low-frequency response is required. A solution to this dilemma that is very practical with LICs is the *differential amplifier* shown in Fig. 11-7. Even though it requires three transistors, they are very cheap in LICs and the effective emitter bypass works at low frequencies,

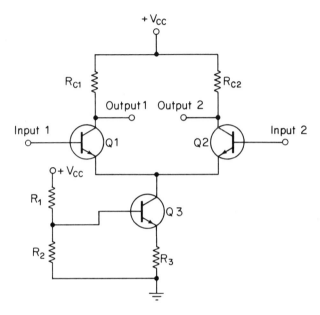

FIG. 11-7. Differential amplifier

even direct current. The $R1$–$R2$ combination biases $Q3$ such that it acts as a constant current source for the emitters of $Q1$ and $Q2$. Hence the emitter-to-ground voltage of $Q1$ and $Q2$ will always remain constant, regardless of the ac signal applied to either input. Therefore, the entire input signal is dropped across the base–emitter junction to which it is applied, and no emitter degeneration takes place at all frequencies down to direct current. Thus a highly temperature stable, high voltage gain amplifier results.

The output may be at either $Q1$'s or $Q2$'s collector. If a noninverting amplifier is needed, the output should be taken at the collector of the transistor opposite to where the input is applied. The voltage gain is one half that of either of the transistors operating at the same dc current levels with no emitter resistance. This is because the total dc emitter current (supplied by $Q3$) is split evenly between $Q1$ and $Q2$ if they have the desired matched characteristics. With one half the dc emitter current the gain is cut in half, since the two are directly proportional. However, if the output is taken differentially across the two collectors, the gain is doubled and is thus back to the same level as with a single transistor amplifier. If signals are applied to both inputs, then the output between the two collectors is proportional to the difference between the imput signals; hence the name differential amplifier. This circuit is the heart of most operational amplifiers, and we shall pursue it in detail later. The importance of this circuit now, however, is that a high voltage gain is now possible in a highly stable amplifier without requiring the emitter bypass capacitor.

EXAMPLE 11-4

Determine the single-ended and differential output for the circuit of Fig. 11-8.

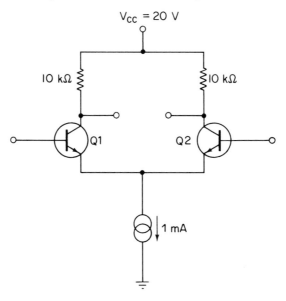

FIG. 11-8. Example 11-4

Solution:

The current source supplies 1 mA of dc current, which is split evenly between $Q1$ and $Q2$. Thus the single-ended gain will be R_C/h_{ib}, where $h_{ib} = 0.026/0.5$ mA $= 52\ \Omega$. Thus

$$G_v \simeq \frac{10\ \text{k}\Omega}{52\ \Omega} \simeq 200$$

for the single-ended output, whereas taking the output differentially across the two collectors results in double that gain, or about 400.

D. Simulation of Inductance

As previously stated, inductors cannot be fabricated in LICs. However, by making use of the *gyrator* principle, inductance can be simulated. Figure

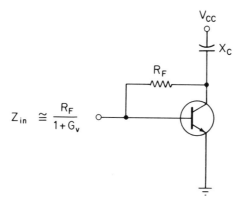

FIG. 11-9. Gyrator

11-9 provides a simple circuit illustrating the gyrator effect. The biasing circuit is omitted. The amplifier's input impedance is approximately

$$Z_{IN} \simeq \frac{R_F}{1 + G_v} \simeq \frac{R_F}{G_v}$$

and its voltage gain is

$$G_v \simeq \frac{R_C}{h_{ib}} = \frac{X_C}{h_{ib}}$$

so that

$$Z_{IN} \simeq \frac{R_F \times h_{ib}}{X_C} \qquad (11\text{-}1)$$

If we let the $R_F \times h_{ib}$ equal a constant, A, then,

$$Z_{IN} = \frac{A}{X_C} = \frac{A}{X \underline{/-90°}} = AB \underline{/+90°}$$

where B is a constant equal to $1/X$. The capacitive load reactance therefore appears as an inductive reactance at the amplifier's input terminals.

EXAMPLE 11-5

Determine the inductance presented by the input terminals of the circuit shown in Fig. 11-10 at a frequency of 10 kHz.

Solution:
At 10 kHz the 1000 pF capacitor has a reactance of $-j\ 15.9\ \Omega$, thus assuring that the combined load impedance with the 10 kΩ resistor is almost purely reactive.

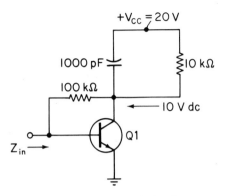

FIG. 11-10. Example 11-5

We can then calculate the effective input reactance from Eq. (11-1) as

$$Z_{IN} \simeq \frac{R_F \times h_{ib}}{X_C}$$

$$= \frac{10^5 \,\Omega \times 26 \,\Omega}{15.9 \,\Omega \,\underline{/-90°}} \qquad (11\text{-}1)$$

$$= 1.64 \times 10^5 \,\Omega \,\underline{/+90°}$$

Thus the input impedance is 164,000 Ω of inductive reactance. Solving for L at 10 kHz,

$$L = \frac{X_L}{2\pi f} = \frac{1.64 \times 10^5}{2\pi \times 10^4}$$

$$= 2.6 \text{ H}$$

11-5 A General-Purpose LIC—The CA3035

The CA3035 LIC was originally developed by Radio Corporation of America as a sense amplifier for TV remote-control circuitry. Its schematic is shown in Fig. 11-11 and it consists of three separate amplifier sections that can be used singly or cascaded. Since each of the three amplifiers is a separate entity, external circuitry must be used to cascade them together. They are all well temperature compensated such that operation over a very wide −55 to +125°C range is possible. When cascaded together, a total voltage gain of 129 dB (2,800,000) is possible. This is more gain that you will likely ever require for any given application.

Because of its low price, it is often an economical replacement for discrete circuitry, even if only one or two of its three sections are utilized. The

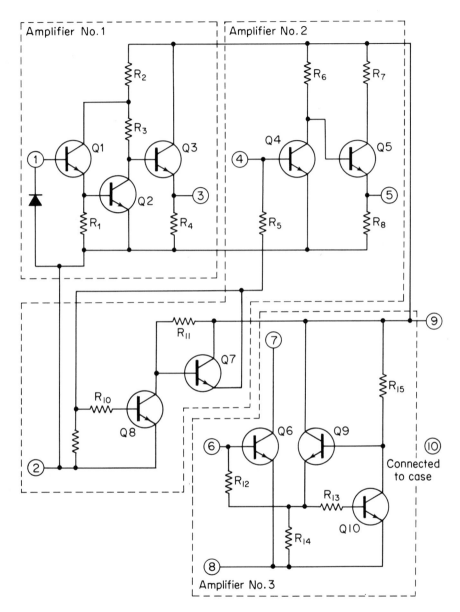

FIG. 11-11. CA3035 LIC schematic (Courtesy of Radio Corporation of America.)

CA3035 sells for under $3.00 in single lots and under $2.00 when 100 or more are purchased. Its extreme versatility is enhanced by its being able to work from a single power supply over a range of about 4–18 V. Each amplifier offers different characteristics, and therefore intelligent application of the device requires a knowledge of each one. Figure 11-12b shows amplifier 1

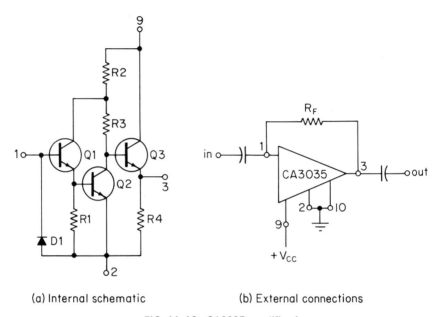

(a) Internal schematic   (b) External connections

FIG. 11-12. CA3035 amplifier 1

with some external circuitry, a coupling capacitor for input and output, and a feedback resistor, R_F. The input at pin 1 is $Q1$, a CC configuration to allow high input impedance with its output driving $Q2$ operating as a high voltage gain CE stage. Notice no emitter resistance is used. Its output drives another emitter follower, $Q3$, to provide a low output impedance. The diode $D1$ protects the amplifier from damage by high input signals. The amplifier offers 50 kΩ input impedance, 270 Ω output impedance, and an open-loop voltage gain of 160. Its low-frequency response is limited only by the external coupling capacitors, and its high-frequency cutoff, f_2, is 500 KHz. Above that the gain falls off at 6 dB per octave such that a gain of 10 is still provided at 6 MHz. This amplifier has a very low noise level and hence is suited for low-level input signals. The output signal should be limited to 1 V p-p if very little distortion can be tolerated; otherwise 2 or 3 V p-p can be obtained. The actual amount of gain can be limited by the value of the external negative-feedback resistor, R_F.

EXAMPLE 11-6

Determine the voltage gain of the amplifier shown in Fig. 11-12 if R_F is omitted, $R_F = 1$ MΩ, and $R_F = 200$ kΩ.

Solution:

With R_F omitted there is no feedback, and hence the gain is simply the open-loop gain or 160.

With a 1MΩ negative feedback resistor feeding back into a 50 kΩ input resistance there is a feedback factor of

$$B = -\frac{50 \text{ k}\Omega}{1 \text{ M}\Omega} = -\frac{1}{20}$$

Recall that for a high open loop gain amplifier

$$G'_v \simeq -\frac{1}{B} \tag{5-5}$$

Therefore,

$$G'_v \simeq \frac{1}{1/20} = 20$$

and when $R_F = 200$ kΩ,

$$G'_v = \frac{1}{50/200} = 4$$

Amplifier 2, shown in Fig. 11-13 with external circuitry, is made up of $Q4$ and $Q5$, with $Q7$ and $Q8$ utilized strictly to apply a highly temperature stabilized base bias to $Q4$ via resistor $R5$. $Q4$ is a high-gain, zero emitter resistance CE amplifier with $Q5$ performing the emitter-follower function to provide low output impedance at pin 5. It has a typical voltage gain of 200, input resistance of 170 Ω, and an extremely high value of f_2 (2.5 MHz). Once again its output should be limited to 1 V p-p for low-distortion applications.

Amplifier 3, shown in Fig. 11-14, is made up simply by one active device, $Q6$, with the other two transistors, $Q9$ and $Q10$, providing a thermally stabilized base bias via $R12$. Its input is at pin 6, and it is to be used with an external load resistor connected between $Q6$'s collector (pin 7) and the positive power supply output. This stage exhibits the following typical characteristics: an input resistance of 670 Ω, voltage gain of 120, $f_2 = 2.5$ MHz, and an output resistance equal to whatever external load resistance is used at $Q6$'s collector. It can handle output signals up to 8 V p-p, but distortion is evident at these high levels.

The three stages can be usefully utilized alone, or cascaded in any combination of two, or with all three cascaded together, as shown in Fig. 11-15.

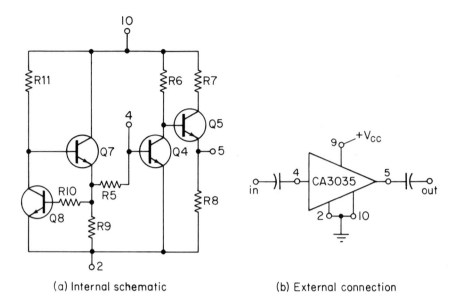

(a) Internal schematic (b) External connection

FIG. 11-13. CA3035 amplifier 2

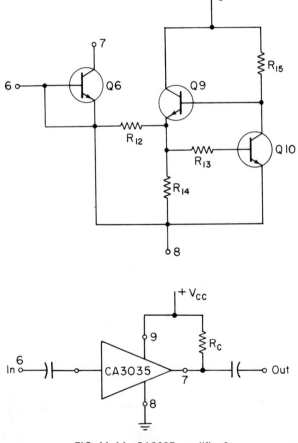

FIG. 11-14. CA3035 amplifier 3

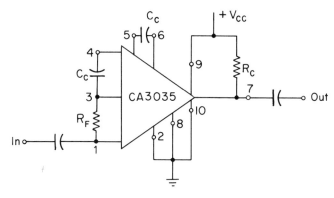

FIG. 11-15. CA3035 with all three amplifiers cascaded

The applications for this device and others similar to it are truly unlimited. Linear integrated circuits such as this should be strongly considered for almost all amplification applications over discrete circuits, since in today's market a distinct economic improvement will often result. Specific applications are not included here, since the reader should now be prepared to adapt this form of circuit to almost any application requiring amplification.

11-6 LIC Power Amplifiers

An interesting development in recent years is the integration of complete power amplifiers into a compact package. They operate class AB for minimum distortion and maximum efficiency. A good example of such a device is shown to scale in Fig. 11-16. Notice that it is labeled "hybrid power IC." This is because all small-signal functions are handled by a LIC while the power output stages and several small capacitors are normal discrete components all encapsulated together with the LIC. The resulting package is compact, reliable, easy to use, and economical. The model shown has a maximum of 50 W ac output into an 8 Ω speaker at a cost below $20.00. A similar unit rated at 10 W is available for less than $5.00 and a 25 W version is also offered. It would be extremely difficult to duplicate them in any form for that money. Table 11-2 provides a schematic diagram as well as other pertinent data for this device.

$Q1$ is the input transistor operating in the CE configuration. The output is fed back to its emitter through pin 3 to provide negative feedback. $Q1$'s output is capacitively coupled to $Q2$, another CE stage, but this time with an emitter bypass capacitor. Its output drives both $Q3$ and $Q4$, which are biased class AB by the diode resistor combination. They in turn drive the output

Table 11-2 Sanken SL-1050A Power Amplifier

TYPICAL CHARACTERISTIC CURVES

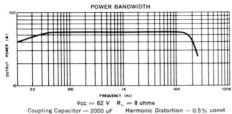

POWER BANDWIDTH
$V_{cc} = 62$ V $R_L = 8$ ohms
Coupling Capacitor = 2000 uF   Harmonic Distortion = 0.5% const.

HARMONIC DISTORTION-OUTPUT POWER
$R_L = 8$ ohms

INPUT VOLTAGE vs. OUTPUT POWER
$R_L = 8$ ohms $f = 1$ KHz

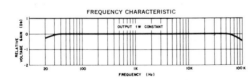

FREQUENCY CHARACTERISTIC
OUTPUT 1W CONSTANT

SCHEMATIC

BOTTOM VIEW

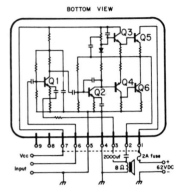

1. Vcc 2. Output (to a capacitor) 3. Feedback 4. Ground for output 5. Ground for input 6. Input 7. Vcc 8. and 9. Spares
*Terminal No. 7 is provided for a ripple filter circuit.

OUTLINE DRAWING
in mm (approx. inches)

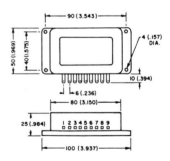

SPECIFICATIONS

Power: Output power ratings are maximum continuous at 1000 Hz with a distortion less than ½%, a 25°C ambient, a load of 8 ohms, and the recommended heat sink and mounting.

Response: Flat within ½ db from 20 Hz to 100,000 Hz, with specified feedback arrangement, as measured at 1 watt output.

Temperature Compensation: An internal compensating diode is used to provide minimum cross-over distortion and protection from thermal runaway.

Voltage Gain: The feedback resistor provided internally allows 30 db gain (typical) when feedback terminal 3 is connected directly to an 8 ohm load. Open loop gain is approximately 62 db.

Special Applications: Parallel and bridge operation for higher power, variations in load, power supply regulation, ambient temperature variations, and variations in feedback are described in Technical Bulletin 70-02 QA.

Heat Sinks: Values shown are minimum for a plain white aluminum sheet, 2 mm (approximately 1/16 inch) thick, at a 25°C ambient, with reasonable ventilation. A silicone grease such as GE Insulgrease G-640 should be used to provide good thermal contact from base to heat sink.

Power Supply: Maximum voltage values are absolute maximum. A transformer with 10% regulation is recommended to assure withstanding a 5 second output short. Idling current increases exponentially with supply voltage, increasing heating.

Derating: Idling current remains constant at any output power level. Internal power loss reaches its maximum when the output is 40% of the rated maximum output and decreases approximately 25% at full power level, if supply voltage does not change. Refer to Bulletin 70-02 QA for complete application data.

Application: (1) Amplifiers may be damaged by oscillation or overdriving. (2) Provide separate ground connections to both input and output. (3) For loads of less than 8 ohms see Technical Bulletin 70-02 QA. (4) Do not exceed recommended power supply voltage. (5) Fuses should be quick-acting type such as Bussman type AGC.

Table 11-2 (cont.)
ELECTRICAL CHARACTERISTICS

Characteristic	SI-1025A	SI-1050A
Maximum rms power	25W	50W
Output Load	8 ohms	8 ohms
Supply Voltage	48V	62V
Absolute Max. Supply Voltage	55V	80V
Supply Current	.8A	1.1A
Suggested Fuse	1A	2A
Harmonic Distortion at Full Output	0.5% max.	0.5% max.
Voltage Gain, Full Feedback	30 db typ.	30 db typ.
Input Impedance	70,000 ohms typ.	70,000 ohms typ.
Output Impedance	0.2 ohms typ.	0.2 ohms typ.
Output Coupling Capacitor	2000 uF 50 WVDC	2000 uF 75 WVDC
Signal to Noise Ratio	90 db typ.	90 db typ.
Idling Current	30 ma typ.	30 ma typ.
Heat Sink (Minimum)	70 cm^2 (11 sq. in.)	135 cm^2 (21 sq. in.)
Operating Temperature	−20°C to +80°C	−20°C to +80°C
Storage Temperature	−30°C to +100°C	−30°C to +100°C

At 25°C ambient, 1 KHz, R_L=8 ohms.

(Courtesy of Airpax Electronics.)

FIG. 11-16. Hybrid LIC power amplifier (Courtesy of Airpax Electronics.)

transistors $Q5$ and $Q6$ and also bias them class AB. The output at pin 2 then drives the speaker through an externally connected 2000 μF coupling capacitor. All that the user need supply besides the coupling capacitor is a 62 V power source at 1.1 A and an input signal of 700 mV to obtain a 50 W low-distortion output.

PROBLEMS

1. Discuss the relative merits of monolithic ICs versus hybrid ICs and conventional printed circuits.
2. Repeat Example 11-1 for a production run of 2000 circuits and 200,000 circuits.
3. Design a dc level shifting circuit similar to the one shown in Fig. 11-4 to shift from a 12 V level to a 2 V level when the power source is 20 V. Specify all resistor values. Assume $h'_{fe} = 100$.
4. Determine the *exact* amount of signal attenuation introduced by the circuit designed in Problem 3. The exact figure is important for a particular application, and therefore you must consider all possible effects.
5. Determine the change to Example 11-3 if the transistors both had h'_{fe}'s

of 50 instead of 250. Would this have any practical effects on the circuit's ability to provide a high-quality, high-valued capacitance?

6. Modify the circuit of Problem 5 such that Z_{IN} is still at least 40 MΩ as in the original example with transistors having an h'_{fe} of 50 being the only ones producible.

7. Repeat Example 11-4 if the current source were decreased to $\frac{1}{2}$ mA. Sketch the differential output if identical 180° out-of-phase sine waves are applied to the bases. Sketch the differential output if identical in-phase sine waves are applied to the bases.

8. Design a gyrator that presents an inductance of 1 H at 4 kHz.

9. Design a gyrator that maximizes the effective inductance presented by a 100 pF capacitor at 5 kHz.

10. Design an amplifier system using the CA3035 LIC to deliver 5 V p-p into a 5 kΩ load from a 5 μV p-p, 200 kΩ source.

11. Explain how transistors $Q7$ and $Q8$ are able to provide a stable base bias to $Q4$ in the CA3035.

12. Explain how transistors $Q9$ and $Q10$ are able to provide a stable base bias to $Q6$ in the CA3035.

13. Which transistors provide voltage gain in the CA3035?

14. Design a Wien bridge oscillator using a CA3035.

15. The CA3035 has good high-frequency response. In some low-frequency applications it is desirable to reduce f_2 to minimize parasitic oscillations. Explain how this could be accomplished with the CA3035.

16. What is the voltage gain of the Sanken S1-1050A hybrid power IC?

17. Calculate the size of output capacitor required for the Sanken S1-1050A if f_1 is to be 10 Hz and $R_L = 8\ \Omega$.

18. Design a preamplifier for the Sanken S1-1050A such that a phono cartridge with 1 mV p-p output and 5 kΩ internal resistance can drive the 8 Ω speaker to the 50 W level.

12

Linear Integrated Circuit Operational Amplifiers

12-1 Introduction

The recent availability of low-cost monolithic integrated circuits (ICs) has revolutionized the field of linear electronics. Whole circuits and subsystems are now available, predesigned, in extremely compact sizes, and with high reliability. As mentioned in Chapter 11, these circuits are replacing many of the discrete designs in today's electronics systems. The most common circuit configuration for LICs is the operational amplifier. An operational amplifier is so named because it originally was used to perform mathematical "operations" in analog computers. However, because of its versatility and the low cost made possible by monolithic manufacturing techniques, its use has spread into virtually every area of electronics. An operational amplifier is a very high gain direct-coupled amplifier that uses external feedback to control its gain–impedance characteristics. The most common circuit form for operational amplifiers is two cascaded differential amplifier stages directly coupled into one or more emitter followers to provide a low output impedance.

The differential amplifier is an ideal circuit for monolithic techniques, since it offers high gains without requiring capacitors or large-sized resistors, requires the closely matched transistor characteristics offered by LICs, and offers high gains determined by resistor ratios instead of absolute values. This circuit can be adapted to virtually any amplification situation, thus allowing high-volume operational amplifier production for the diverse linear circuit applications. This high-volume production, you may recall, is a requirement for making monolithic circuits economically attractive.

Operational amplifier circuits intended primarily for narrowband RF and IF applications can be manufactured with the collectors of a differential pair of transistors uncommitted. In this way the user may apply the appropriate tuned circuit externally to the LIC package. For most other circuits collector resistors are a part of the package, and the user can tailor the operational amplifier to his needs through judicious use of the feedback and phase compensation elements (which are usually externally added) and sometimes an external bias control.

This chapter provides the basic information that will allow the reader to intelligently deal with this truly powerful electronic circuit element. Much of the material that has been developed throughout this text is utilized to present this circuit, and it, in a sense, is a culmination of your linear circuits study—a stepping stone into complete communication systems.

12-2 Ideal Operational Amplifier

In its simplest form, an operational amplifier is represented by a triangle with two inputs and one output, as shown in Fig. 12-1. It does not show any

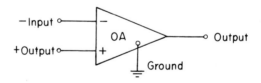

FIG. 12-1. Operational amplifier symbol (simplified)

of the other necessary connections, such as for dc power, feedback connections, ground, or phase compensation. The "(−) input" is termed the *inverting* input since signals applied between it and ground appear inverted at the output (with respect to ground). For similar (but inverse) reasons the "(+) input" is referred to as the *noninverting* input. The input may also be applied *differentially*, i.e., between the two inputs instead of to just one of them with respect to ground.

The ideal operational amplifier has the following characteristics:
1. Infinite gain (open loop).
2. Infinite input impedance (open loop).

3. Infinite bandwidth.
4. Zero output impedance.

These characteristics are obviously not possible, but in practice the first two are so closely attained that analysis can be made by using them as approximations. This greatly simplifies development of operational amplifier gain–impedance relationships without adversely affecting the results obtained. Therefore, we shall assume practical operational amplifiers with infinite gain and infinite input impedance.

Let us first analyze the inverting operational amplifier shown in Fig. 12-2. Notice that the positive input is grounded and that resistor $R2$ provides

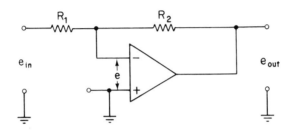

FIG. 12-2. Inverting operational amplifier

feedback from output to input. If we let e be the voltage from the inverting input to ground, we can write the following relationship:

$$\frac{e_{in} - e}{R1} = \frac{e - e_{out}}{R2} \qquad (12\text{-}1)$$

Equation (12-1) is valid since the operational amplifier's input impedance is ideally infinite looking into the inverting input, and thus no current flows into the amplifier. Hence the current through $R1$ must equal the current through $R2$. Now since the open-loop gain of the operational amplifier is ideally infinite, it follows that the voltage e is zero; therefore, Eq. (12-1) reduces to

$$\frac{e_{out}}{e_{in}} = -\frac{R2}{R1} \qquad (12\text{-}2)$$

This is the *closed-loop gain* of the amplifier (with feedback), and we shall hereafter refer to it as A_F. Thus

$$A_F = -\frac{R2}{R1} \qquad (12\text{-}3)$$

We shall call the amplifier's *open-loop gain* A_o. We have assumed A_o to be infinite in this derivation and throughout the chapter.

The same results could have been obtained by using the general feedback relationships developed in Chapter 5. In that format

$$A_F = \frac{A_o}{1 + A_o B} \qquad (12\text{-}4)$$

and B is the feedback factor, which is

$$B = -\frac{R1}{R2} \qquad (12\text{-}5)$$

such that

$$A_F = \frac{A_o}{1 + A_o[-(R1/R2)]}$$

If A_o is infinite,

$$A_F = \frac{1}{-(R1/R2)} = -\frac{R2}{R1}$$

as in Eq. (12-3).

The input impedance for the circuit of Fig. 12-2 will equal the input voltage divided by the input current:

$$R_{IN} = \frac{e_{in}}{(e_{in} - e)/R1} \qquad (12\text{-}6)$$

Assuming that the operational amplifier input impedance is infinite, this will reduce to

$$R_{IN} = R1 \qquad (12\text{-}7)$$

Thus we have so far discussed the input impedance, open-loop gain (A_o), and the closed-loop gain (A_F). The *loop gain* (L.G.) is also of interest and is defined as the product of the open-loop gain and the feedback factor. Thus

$$\text{loop gain} = \text{L.G.} = A_o B \qquad (12\text{-}8)$$

and it is also equal to A_o/A_{FB}, since A_{FB} equals $1/B$.

The noninverting version of the operational amplifier is shown in Fig. 12-3. Note that the ($-$) terminal is not grounded and that $R2$ and $R1$ are still connected as in the inverting mode. Since the ideal gain of the operational amplifier itself is infinite, it seems logical that the voltage between the inverting and noninverting terminals is negligible, and hence the voltage across $R1$ should equal e_{in}. Also, since no current will ideally be drawn by the inverting input, the currents through $R2$ and $R1$ are equal. Thus

$$\frac{e_{out} - e_{in}}{R2} = \frac{e_{in}}{R1} \qquad (12\text{-}9)$$

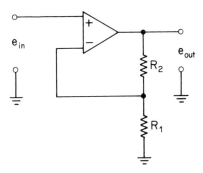

FIG. 12-3. Noninverting amplifier

and this can be manipulated to give a closed-loop gain of

$$A_F = \frac{e_{\text{out}}}{e_{\text{in}}} = 1 + \frac{R2}{R1} \quad (12\text{-}10)$$

Thus the gain is positive and once again is determined by two resistors that are externally added to the operational amplifier package. The input impedance of the ideal noninverting circuit is infinite. In practice, however, some finite high value of input resistance is present. The operational amplifier manufacturer provides this as the open-loop input resistance, and it is usually given the symbol Z_{in}. In the feedback mode of operation this impedance is multiplied by the ratio of A_o to A_{FB}. Thus the closed-loop circuit input impedance is

$$R_{\text{IN}} = Z_{\text{IN}} \frac{A_o}{A_{FB}} = \frac{Z_{\text{IN}} A_o}{1 + (R2/R1)} \quad (12\text{-}11)$$

Since A_o/A_{FB} is equal to the loop gain, Eq. (12-11) reduces to

$$R_{\text{IN}} = Z_{\text{IN}} \times \text{L.G.} \quad (12\text{-}12)$$

The noninverting circuit offers a higher input impedance than its inverting circuit counterpart.

Table 12-1 summarizes these results for the two forms of operational amplifiers. We have not been concerned with their output impedances. However, virtually all operational amplifiers can drive into a 500 Ω load without appreciable loading, but to do significantly better would require the user to add his own power amplifier. The operational amplifier manufacturers do not normally provide output impedance information except indirectly through a maximum output power rating.

Table 12-1 Operational-Amplifier Summary

	INVERTING CIRCUIT	NONINVERTING CIRCUIT
Closed-loop voltage gain, A_F	$-\dfrac{R2}{R1}$	$1 + \dfrac{R2}{R1} = \dfrac{R1 + R2}{R1}$
Loop gain, L.G. $(= -A_o B)$	$\dfrac{-A_o R1}{R2}$	$\dfrac{A_o R2}{R1 + R2}$
R_{IN}	$R1$	$Z_{IN} \times$ L.G.

12-3 Applications

Even though we have considered the operational amplifier mainly from an ideal standpoint up to this point, it is useful to consider some of their applications before further study of the device. In fact, many scientific people without any electronics background have learned to apply these devices in their work—treating the operational amplifier as a "black box" that when properly connected can be set to useful work. Naturally, a person such as yourself can apply the device more effectively, however, because of your electronic knowledge and your understanding of the inner goings on of the operational amplifier itself.

A. Small-Signal Amplifiers

The first application we shall consider for the operational amplifier is that of a replacement for a discrete-component small-signal amplifier. Consider the three-stage amplifier we analyzed in Section 9-9 (Fig. 9-8). That circuit, with its three transistors, three capacitors, and nine resistors, offered a voltage gain of 360 over a 40–83 kHz bandwidth with a 12.1 kΩ input impedance and an output impedance of 40 Ω. An operational amplifier and two resistors could very easily exceed all those specifications, except perhaps the output impedance, and in much less space and for much less money. Commercial-grade operational amplifiers with specifications more than adequate for this purpose are available for well under $1.00, which makes it ludicrous to use the discrete circuits in modern electronics. Figure 12-4 shows an operational amplifier circuit providing a gain of 360 from direct current on up, with an input impedance of at least 100 kΩ. Depending on the operational amplifier used, the output impedance of 40 Ω may or may not be obtained. However, if not, the addition of a single emitter follower would clear up that problem. One glance at the circuit (Fig. 9-8) replaced by Fig. 12-4 (at probably one fifth the cost) should provide the reader with an indication of the opera-

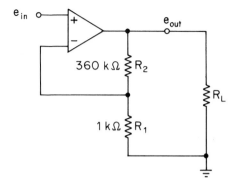

FIG. 12-4. Small-signal amplifier using operational amplifier

tional amplifier's great value. Notice that the noninverting configuration was used here, because the inverting circuit would have yielded an unacceptably low input impedance.

B. Follower Circuit

An operational amplifier is also used as a follower circuit, as in Fig. 12-5, whenever extremely high input impedance and low output impedance

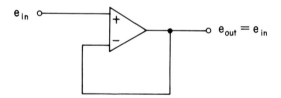

FIG. 12-5. Operational amplifier follower circuit

are required. The input impedance is the circuit's open-loop gain A_o plus 1 multiplied by the unit's open-loop input impedance Z_{IN}:

$$R_{IN} = Z_{IN}(1 + A_o) \qquad (12\text{-}13)$$

Notice that it is the unit's input impedance that is multiplied, which then implies that the input capacity is divided, since $X_C = 1/(2\pi f C)$. The follower circuit therefore finds use in very high frequency applications, as well as for large impedance transformation applications.

C. Comparator

A comparator is, simply stated, a circuit that compares two signals. They are widely used for level detection. Shown in Fig. 12-6, the comparator is the

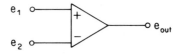

FIG. 12-6. Comparator

simplest operational amplifier circuit with no additional external components necessary. The circuit works with full open-loop gain. If e_1 and e_2 are equal, then e_{out} should ideally be zero. If e_1 should change by even a very small amount from e_2's value, e_{out} should increase by a large amount because of the large value of open-loop gain, A_o. Because of the high gain involved, the output is usually "saturated" at the full positive or full negative level. Thus a rectangular-type waveform is the normal output, regardless of the waveforms being compared. The circuit thus detects small level changes, which is another way of saying it compares two signals.

D. Adder

The adder circuit provides an output either equal to or proportional to the sum of two or more signals. By prudently choosing resistor values, it can be used to provide an output equal to the average value of the inputs. Figure 12-7 shows a three-input adder circuit. All three inputs are applied to the

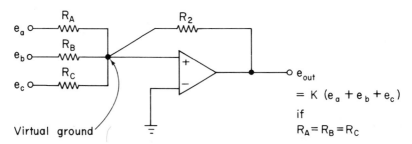

FIG. 12-7. Three-input adder circuit

inverting input through separate resistors. Since the voltage at the noninverting input is virtually the same as at the inverting input (due to the high open-loop gain), we can say that the inverting input is "almost" grounded. It is known as a *virtual* ground and results in each source voltage seeing only its respective summing resistor. The sum of the three input currents must equal the current through $R2$, since our ideal operational amplifier draws no current itself. Therefore,

$$\frac{e_a}{R_A} + \frac{e_b}{R_B} + \frac{e_c}{R_C} = -\frac{e_{out}}{R2} \quad (12\text{-}14)$$

and if we let $R_A = R_B = R_C = R$, Eq. (12-14) can be simplified to

$$e_{\text{out}} = -\frac{R2}{R}(e_a + e_b + e_c) \tag{12-15}$$

Thus the output is proportional to the sum of the inputs. If $R2 = R$, the output exactly equals the *sum* of the inputs. If $R2 = 3R$, the output would equal the sum of the inputs divided by 3 or an output equal to the average of the input signals.

E. Subtractor

As would be expected, the subtractor circuit shown in Fig. 12-8 provides an output equal to the difference of two signals. Since neither operational

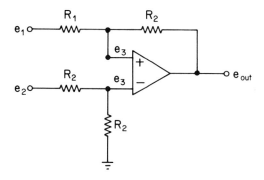

FIG. 12-8. Subtractor circuit

amplifier input is at ground potential, we shall assume an input potential, with respect to ground, of e_3 for each one. Remember that the potentials at each input will be almost equal, since the operational amplifier itself has such a high loop gain. Now recalling that the operational amplifier inputs effectively draw zero current,

$$\frac{e_1 - e_B}{R1} = \frac{e_B - e_{\text{out}}}{R2} \tag{12-16}$$

and

$$\frac{e_2 - e_B}{R1} = \frac{e_B}{R2} \tag{12-17}$$

By eliminating e_3 from these two equations and solving for e_{out},

$$e_{\text{out}} = \frac{R2}{R1}(e_2 - e_1) \tag{12-18}$$

If $R2 = R1$, then e_{out} simply equals the difference between e_2 and e_1. The input impedance seen by e_2 is $R1 + R2$, but unfortunately the impedance seen by e_1 is voltage dependent, being predicted by

$$\frac{e_1}{e_2} = \frac{e_1 R1}{(e_1 - e_2)/2} \tag{12-19}$$

Thus in high-accuracy applications it is necessary to drive the subtractor from a relatively low impedance source.

F. Integrator

There is considerable demand for a circuit that provides the mathematical integral of a given signal. The most simple example of integration is shown in Fig. 12-9 where the input signal is a dc level and the integral is a linearly increasing ramp output. Applying a dc level to an RC circuit, as shown in Fig. 12-10a, results in an integrator-type output initially, but as the voltage

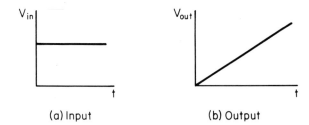

FIG. 12-9. Input and output signals for an integrator

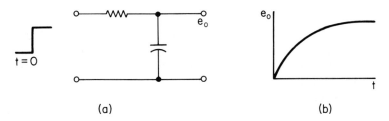

FIG. 12-10. RC integrator and output signal

charge across the capacitor builds up, its charging current decays exponentially because the voltage across R is decreasing. An operational amplifier circuit such as shown in Fig. 12-11 can supply the constant current necessary to allow a highly linear charge current to the capacitor, which results in a *highly linear* output. Applying a dc input causes the voltage across R to be constant, since the inverting input is a virtual ground (as shown in dashed lines in Fig. 12-11). Thus the current through R is constant, and it is supplied only to C, since the operational amplifier itself draws negligible current. The

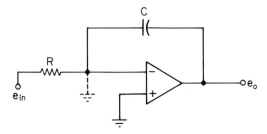

FIG. 12-11. Operational amplifier integrator

voltage across C therefore increases at a linear rate and is equal to the negative value of e_o. The operational amplifier provides a method of generating highly accurate ramp voltages, as may be required to drive cathode-ray tubes, and also can provide highly accurate integrals of any other signal.

G. Differentiators

By interchanging the resistor and capacitor of the integrator circuit, a differentiator circuit is formed. Differentiation is a mathematical operation that provides the rate of change of a signal. Thus feeding a linear ramp that has a constant rate of change into a differentiator yields a constant dc output, as shown in Fig. 12-12. Unfortunately, the operational amplifier differentiator circuit shown in Fig. 12-13 has a gain that increases with frequency, making it highly susceptible to high-frequency noise. A solution to this problem is to add some series resistance to the input capacitor (sometimes accomplished by the source impedance of the input signal) so that the high-frequency gain is reduced.

FIG. 12-12. Differentiator

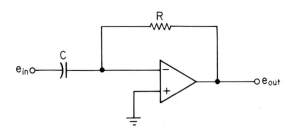

FIG. 12-13. Differentiator

H. Logarithmic Amplifier

The circuit of Fig. 12-14 provides an output that is directly proportional to the natural logarithm of the source voltage. Notice that the transistor's base

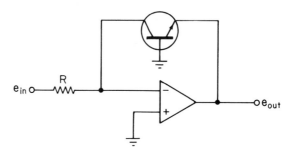

FIG. 12-14. Logarithmic amplifier

is grounded and that its collector is connected to the operational amplifier's inverting input, which is a virtual ground. Hence its collector–base voltage is virtually zero. It is known that a transistor's emitter–base voltage at the $V_{CB} = 0$ condition is predicted by

$$V_{EB} = \frac{1}{Q}\left[\ln\left(\frac{I_C}{P}\right)\right] \tag{12-20}$$

where P and Q are constants depending on the transistor's characteristics and temperature. The output voltage, e_o, is in parallel with, and therefore equal to, V_{EB}; thus

$$e_o = \frac{1}{Q}\left[\ln\left(\frac{I_C}{P}\right)\right] \tag{12-21}$$

The transistor's collector current is also equal to the input current (e_{in}/R), and therefore

$$e_o = \frac{1}{Q}\left[\ln\left(\frac{e_{in}}{RP}\right)\right] \tag{12-22}$$

12-4 Internal Circuitry

We have thus far studied the operational amplifier and some of its applications without any clear idea as to exactly what this device contains. Most operational amplifiers contain a high input impedance differential amplifier that provides high gain, another stage of gain that is usually another

Internal Circuitry **285**

differential amplifier, and an impedance transformation output stage—often some emitter followers and/or a class AB push–pull stage. Figure 12-15 provides a schematic used by a whole series of RCA operational amplifiers.

Transistors $Q1$ and $Q2$ in Fig. 12-15 comprise the differential amplifier for the noninverting and inverting inputs. Transistor $Q6$ provides a constant

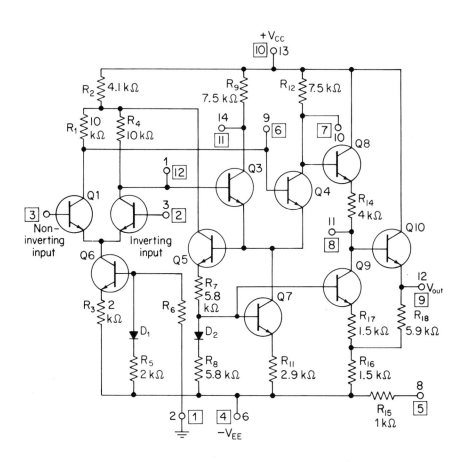

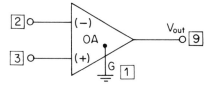

FIG. 12-15. Operational amplifier schematic (Courtesy of Radio Corporation of America.)

current to the emitters of $Q1$ and $Q2$ to allow for proper differential amplifier operation. The diode $D1$ is used to temperature compensate $Q6$ so that the current it supplies is constant over a wide temperature range. Resistors $R5$ and $R6$ form a voltage divider bias for $Q6$ with pin 1 (we shall use the pin numbers enclosed by a square) usually grounded. It may, however, be connected to some variable voltage that is used to control the open-loop gain of this operational amplifier. Varying the voltage applied to pin 1 would vary the amount of dc emitter current to $Q1$ and $Q2$, and hence vary their voltage gain. Recall that $G_v = R_C/h_{ib}$ where $h_{ib} = 0.026/I_E$.

The outputs at the collectors of $Q1$ and $Q2$ drive the second differential stage made up of $Q3$ and $Q4$. Bias stabilization for this stage is provided by constant current transistor $Q7$, which is compensated itself by diode $D2$. The compensating diode, $D2$, also provides thermal stabilization for $Q9$. Transistor $Q5$ is used to keep the output signal as close to zero as possible when the same signal is applied to both inputs of the operational amplifier. Under those conditions, the signal at $Q5$'s base should be zero. If it is not, it develops a signal across $R2$ through $Q5$'s amplification in the proper phase to reduce the error. This same signal is reflected into $Q7$'s base from $Q5$'s emitter to further reduce the error. Transistor $Q5$ also serves to reduce the error in output signal from zero when the two inputs are at the same level, due to drift by either the + or − power supply.

The output of the second differential amplifier stage is taken at $Q4$'s collector and then fed into an impedance-transforming dc-level-shifting network comprised of $Q8$, $Q9$, and $Q10$. Since the output (pin 9) is to be at ground potential with no input signals, it is the function of $Q9$ to supply a constant current for biasing $Q8$ such that the resulting dc level at $Q10$'s emitter is as close to ground as possible, regardless of supply voltage variations and imbalances. Transistors $Q8$ and $Q10$ also provide a low output impedance, being two cascaded emitter followers.

12-5 Frequency Considerations

Figure 12-15 is typical for an entire line of RCA operational amplifiers, including their model numbers CA3008, CA3010, CA3015, CA3016, CA-3029, CA3030, CA3037, and CA3038. These units have a typical open-loop voltage gain (A_o) of 60 dB with a frequency response as shown in Fig 12-16. Notice that the high-frequency cutoff is about 0.2 MHz for the uncompensated amplifier, whereas the compensated curve results when series RC circuits are connected between the collectors of the first differential amplifier to the bases of the second differential amplifier. A typical circuit connection using this form of phase compensation is shown in Fig. 12-17. The external 5 pF capacitors

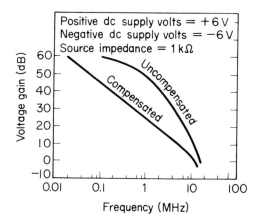

FIG. 12-16. Operational amplifier open-loop gain as a function of frequency (Courtesy of Radio Corporation of America.)

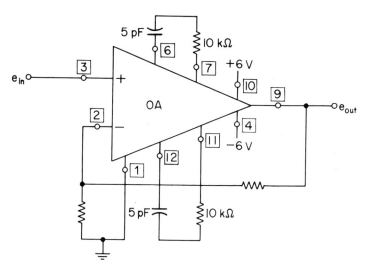

FIG. 12-17. Operational amplifier phase compensation

and 10 kΩ resistors are used to keep the loop phase shift from being equal to 360°. The *RC* circuits thus serve to prevent the high-frequency oscillations that would certainly occur without them. Although some of the newer monolithic operational amplifiers do not require external phase compensation, those that do will not function properly without it. The manufacturer supplies the necessary information on compensation techniques in the operational amplifier data sheet. Notice that the compensation circuits serve to reduce the usable bandwidth of the amplifier. If this operational amplifier were to be used at direct current or only low frequencies, a simple shunt capacitance of appropriate value at any point in the gain loop would suffice to attenuate the high frequencies that would otherwise cause parasitic oscillations.

One other operational amplifier characteristic of interest at high frequencies is known as its slew rate. The *slew rate* is the internally limited rate of change in output voltage with a large-amplitude step function applied to the input. In other words, limitations exist that will not allow the output to change as rapidly as the bandwidth information would lead one to believe for large-signal operation. The slewing rate then is the maximum rate at which the output can change in response to a step input, and is usually specified in volts per microsecond. Although the slew rate and bandwidth both limit the usable frequency range of an operational amplifier, they produce effects that are different in a subtle way. The high-frequency cutoff or bandwidth is a frequency where the sinusoidal output is getting smaller, although it can still be purely sinusoidal. Exceeding the slew rate, however, would not necessarily result in a reduction of the output level, but does result in distortion of the sine wave.

Recall the equation of a sine wave as

$$V = V_P \sin(2\pi ft) \qquad (12\text{-}23)$$

The maximum rate of change for a sine wave occurs as it passes through zero, and this rate of change is

$$\left.\frac{\Delta V}{\Delta t}\right|_{t=0} = 2\pi f V_P \qquad (12\text{-}24)$$

The term $2\pi f V_P$ in Eq. (12-24) should not exceed the manufacturer's published slew rate if an undistorted sinusoidal output is desired.

EXAMPLE 12-1

The bandwidth of a compensated operational amplifier is measured as 500 kHz. The slew rate is given as 1 V/μs. Determine the maximum amplitude output at its high-frequency cutoff.

Solution:
Setting the slew rate equal to $2\pi f V_P$ at 500 kHz yields

$$1 \text{ V}/\mu\text{s} = 2 \times 500 \text{ kHz} \times V_P = \frac{1 \text{ V}}{10^{-6} \text{ s}}$$

Rearranging and solving for V_P, we obtain

$$V_P = \frac{10^6}{2\pi \times 0.5 \times 10^6} = \frac{1}{\pi} = 0.318 \text{ V}$$

Thus at 500 kHz a pure sine wave output can be obtained only up to an output level of 0.318 V peak. At 250 kHz (one half the original frequency), we could obtain twice the undistorted output signal, and so on.

12-6 Common-Mode Effects

One very useful feature of an operational amplifier is the fact that ideally it will provide a 0 V output for zero or equal inputs. Unfortunately, the practical operational amplifier cannot meet the ideal specification but has an output of 1–10 mV under these conditions. In addition, this *offset*, as it is termed, varies with temperature and supply voltage.

Since we are talking about very small voltage levels (<10 mV), it is only in the most critical applications that this operational amplifier characteristic becomes a factor. It is specified on manufacturers' data sheets in several fashions. The *common-mode voltage gain*, A_c, is the ratio of the small output offset signal change to the change in common-mode input when that input signal changes to cause a zero output. This gain is much less than 1. The *common-mode rejection ratio* (CMRR) is the ratio of the common-mode gain (A_c) to the open-loop voltage gain (A_o):

$$\text{CMRR} = \frac{A_c}{A_o} \tag{12-25}$$

It is most frequently expressed in decibels as

$$\text{CMRR(dB)} = 20 \log \frac{A_c}{A_o} \tag{12-26}$$

Typical operational amplifier values for the CMRR are 70–100 dB down. The CMRR is a measure of the ability of the differential amplifier to discriminate between differential and common-mode input signals.

Operational amplifiers are usually given an input offset voltage specification. The *input offset voltage* is the voltage that must be applied between the input terminals to obtain zero output voltage. This voltage varies with temperature, but in critical applications an adjustable resistive divider is set up to null the operational amplifier for one specific temperature. Some operational amplifiers provide two specific terminals for offset nulling. A potentiometer is connected across these two terminals with the wiper usually connected to the negative supply. The *input offset current* is sometimes specified, and it is the difference in the two input currents necessary to provide 0 V output.

One last offset definition to consider is the supply voltage rejection ratio. The *supply voltage rejection ratio* is the ratio of the change in input offset voltage to the change in the supply voltage producing it. This specification allows the designer to choose a power supply with adequate regulation and stability for a given application.

12-7 The µA741 Operational Amplifier

In this section, the manufacturer's complete specifications for a modern operational amplifier are presented and discussed. The student will learn much about operational amplifiers by studying this information as well, thus becoming familiar with one of the easiest to use and most versatile monolithic operational amplifiers developed. It has specifications that compare to some of the best hybrid operational amplifiers of 1965 that cost $100, and yet this device is priced below $0.50 in very large quantities and at under $5.00 in single quantities. There can be little question as to why devices such as this are changing the character of the entire linear electronics industry.

Table 12-2 presents the complete specifications and a number of applications for the Fairchild µA741 operational amplifier. It is one of the second-generation monolithic operational amplifiers, because it offers considerable improvement over most characteristics of the initially offered monolithic units. The problems of the first-generation units, which the µA741 overcomes, include

1. Complicated frequency stabilization networks.
2. Lack of output short circuit protection.
3. Low allowable differential input voltages.
4. Lack of simple offset null method.
5. Instability with capacitive loads.

Referring to the equivalent schematic on page 291, several "different" looking configurations are apparent. The unit is essentially a two-stage amplifier comprising a high gain differential input stage, followed by a high gain driver with a class AB output. In the input stages ($Q1$, $Q2$, $Q3$, $Q4$) notice the use of *npn–pnp* combinations. Since high h_{fe} *pnp*s involve costly additional processing steps, the input uses a combination of high h_{fe} *npn*s and low h_{fe} *pnp*s to achieve high input impedances and gain and low input bias currents. To obtain high gain the collectors of $Q5$ and $Q6$ are used as loads for the differential stages, giving effective values of about 2 MΩ. The constant current source and bias stabilization are made up of $Q7$, $Q8$, $Q9$, and $Q10$.

In cases when input offset voltage control is required, an external 10 kΩ potentiometer should be connected between the emitters of $Q5$ and $Q6$ with the wiper connected to the negative supply. Using this technique, the input offset can be adjusted approximately ±25 mV from its initial value.

The output from the input differential stage is then taken from the collector of $Q4$ and fed into a Darlington stage ($Q16$, $Q17$) to avoid loading. It then feeds a complementary symmetry network with $Q14$ and $Q20$ operating class AB with about 60 μA of quiescent bias current to eliminate crossover distortion.

Table 12-2 Fairchild μA741 Operational Amplifier

FEATURES:
- NO FREQUENCY COMPENSATION REQUIRED
- SHORT-CIRCUIT PROTECTION
- OFFSET VOLTAGE NULL CAPABILITY
- LARGE COMMON-MODE AND DIFFERENTIAL VOLTAGE RANGES
- LOW POWER CONSUMPTION
- NO LATCH UP

GENERAL DESCRIPTION — The μA741 is a high performance monolithic operational amplifier constructed on a single silicon chip, using the Fairchild Planar® epitaxial process. It is intended for a wide range of analog applications. High common mode voltage range and absence of "latch-up" tendencies make the μA741 ideal for use as a voltage follower. The high gain and wide range of operating voltage provides superior performance in integrator, summing amplifier, and general feedback applications. The μA741 is short-circuit protected, has the same pin configuration as the popular μA709 operational amplifier, but requires no external components for frequency compensation. The internal 6dB/octave roll-off insures stability in closed loop applications.

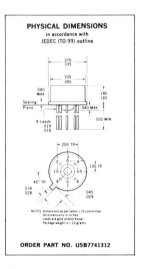

PHYSICAL DIMENSIONS
in accordance with
JEDEC (TO-99) outline

ORDER PART NO. U5B7741312

ABSOLUTE MAXIMUM RATINGS

Supply Voltage	±22 V
Internal Power Dissipation (Note 1)	500 mW
Differential Input Voltage	±30 V
Input Voltage (Note 2)	±15 V
Voltage between Offset Null and V⁻	±0.5 V
Storage Temperature Range	−65°C to +150°C
Operating Temperature Range	−55°C to +125°C
Lead Temperature (Soldering, 60 sec)	300°C
Output Short-Circuit Duration (Note 3)	Indefinite

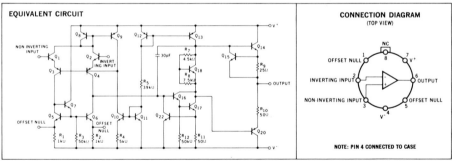

NOTES:
(1) Rating applies for case temperatures to 125°C; derate linearly at 6.5 mW/°C for ambient temperatures above +75°C.
(2) For supply voltages less than ±15 V, the absolute maximum input voltage is equal to the supply voltage.
(3) Short circuit may be to ground or either supply. Rating applies to +125°C case temperature or +75°C ambient temperature.

*Planar is a patented Fairchild process.

Table 12-2 (cont.)

FAIRCHILD LINEAR INTEGRATED CIRCUITS μA741

ELECTRICAL CHARACTERISTICS ($V_S = \pm 15$ V, $T_A = 25°C$ unless otherwise specified)

PARAMETERS (see definitions)	CONDITIONS	MIN.	TYP.	MAX.	UNITS
Input Offset Voltage	$R_S \leq 10$ kΩ		1.0	5.0	mV
Input Offset Current			20	200	nA
Input Bias Current			80	500	nA
Input Resistance		0.3	2.0		MΩ
Input Capacitance			1.4		pF
Offset Voltage Adjustment Range			± 15		mV
Large-Signal Voltage Gain	$R_L \geq 2$ kΩ, $V_{out} = \pm 10$ V	50,000	200,000		
Output Resistance			75		Ω
Output Short-Circuit Current			25		mA
Supply Current			1.7	2.8	mA
Power Consumption			50	85	mW
Transient Response (unity gain)	$V_{in} = 20$ mV, $R_L = 2$ kΩ, $C_L \leq 100$ pF				
Risetime			0.3		μs
Overshoot			5.0		%
Slew Rate	$R_L \geq 2$ kΩ		0.5		V/μs
The following specifications apply for $-55°C \leq T_A \leq +125°C$:					
Input Offset Voltage	$R_S \leq 10$ kΩ		1.0	6.0	mV
Input Offset Current	$T_A = +125°C$		7.0	200	nA
	$T_A = -55°C$		85	500	nA
Input Bias Current	$T_A = +125°C$		0.03	0.5	μA
	$T_A = -55°C$		0.3	1.5	μA
Input Voltage Range		± 12	± 13		V
Common Mode Rejection Ratio	$R_S \leq 10$ kΩ	70	90		dB
Supply Voltage Rejection Ratio	$R_S \leq 10$ kΩ		30	150	μV/V
Large-Signal Voltage Gain	$R_L \geq 2$ kΩ, $V_{out} = \pm 10$ V	25,000			
Output Voltage Swing	$R_L \geq 10$ kΩ	± 12	± 14		V
	$R_L \geq 2$ kΩ	± 10	± 13		V
Supply Current	$T_A = +125°C$		1.5	2.5	mA
	$T_A = -55°C$		2.0	3.3	mA
Power Consumption	$T_A = +125°C$		45	75	mW
	$T_A = -55°C$		60	100	mW

TYPICAL PERFORMANCE CURVES

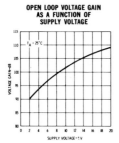

OPEN LOOP VOLTAGE GAIN AS A FUNCTION OF SUPPLY VOLTAGE

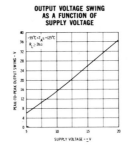

OUTPUT VOLTAGE SWING AS A FUNCTION OF SUPPLY VOLTAGE

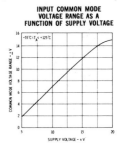

INPUT COMMON MODE VOLTAGE RANGE AS A FUNCTION OF SUPPLY VOLTAGE

Table 12-2 (cont.)

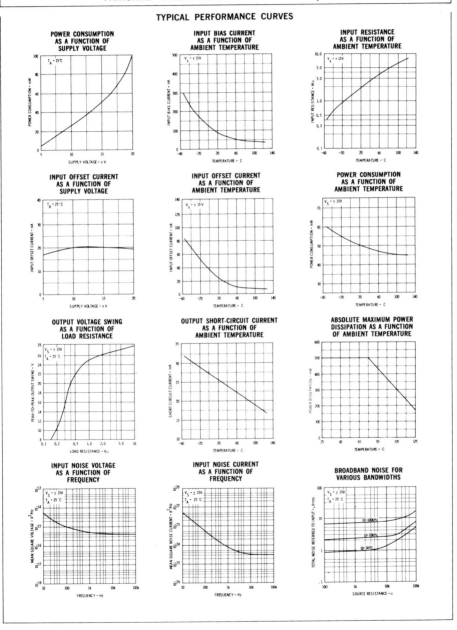

Table 12-2 (cont.)

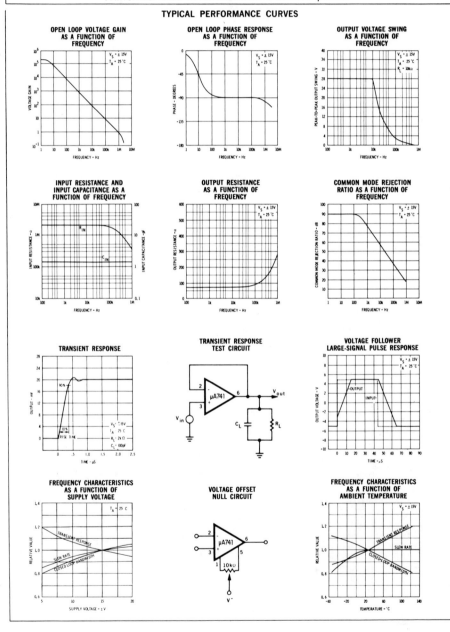

Table 12-2 (cont.)

FAIRCHILD LINEAR INTEGRATED CIRCUITS μA741

DEFINITION OF TERMS

INPUT OFFSET VOLTAGE — That voltage which must be applied between the input terminals to obtain zero output voltage. The input offset voltage may also be defined for the case where two equal resistances are inserted in series with the input leads.
INPUT OFFSET CURRENT — The difference in the currents into the two input terminals with the output at zero volts.
INPUT BIAS CURRENT — The average of the two input currents.
INPUT RESISTANCE — The resistance looking into either input terminal with the other grounded.
INPUT CAPACITANCE — The capacitance looking into either input terminal with the other grounded.
LARGE-SIGNAL VOLTAGE GAIN — The ratio of the maximum output voltage swing with load to the change in input voltage required to drive the output from zero to this voltage.
OUTPUT RESISTANCE — The resistance seen looking into the output terminal with the output at null. This parameter is defined only under small signal conditions at frequencies above a few hundred cycles to eliminate the influence of drift and thermal feedback.
OUTPUT SHORT-CIRCUIT CURRENT — The maximum output current available from the amplifier with the output shorted to ground or to either supply.
SUPPLY CURRENT — The DC current from the supplies required to operate the amplifier with the output at zero and with no load current.
POWER CONSUMPTION — The DC power required to operate the amplifier with the output at zero and with no load current.
TRANSIENT RESPONSE — The closed-loop step-function response of the amplifier under small-signal conditions.
INPUT VOLTAGE RANGE — The range of voltage which, if exceeded on either input terminal, could cause the amplifier to cease functioning properly.
INPUT COMMON MODE REJECTION RATIO — The ratio of the input voltage range to the maximum change in input offset voltage over this range.
SUPPLY VOLTAGE REJECTION RATIO — The ratio of the change in input offset voltage to the change in supply voltage producing it.
OUTPUT VOLTAGE SWING — The peak output swing, referred to zero, that can be obtained without clipping.

TYPICAL APPLICATIONS

UNITY-GAIN VOLTAGE FOLLOWER

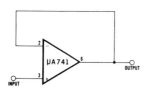

$R_{IN} = 400\ M\Omega$
$C_{IN} = 1\ pF$
$R_{out} \ll 1\ \Omega$
B.W. = 1 MHz

NON-INVERTING AMPLIFIER

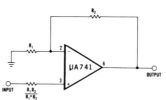

GAIN	R_1	R_2	B.W.	R_{IN}
10	1 kΩ	9 kΩ	100 kHz	400 MΩ
100	100 Ω	9.9 kΩ	10 kHz	280 MΩ
1000	100 Ω	99.9 kΩ	1 kHz	80 MΩ

INVERTING AMPLIFIER

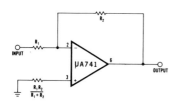

GAIN	R_1	R_2	B.W.	R_{IN}
1	10 kΩ	10 kΩ	1 MHz	10 kΩ
10	1 kΩ	10 kΩ	100 kHz	1 kΩ
100	1 kΩ	100 kΩ	10 kHz	1 kΩ
1000	100 Ω	100 kΩ	1 kHz	100 Ω

CLIPPING AMPLIFIER

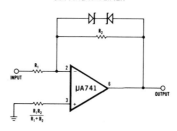

$$\frac{E_{out}}{E_{IN}} = \frac{R_2}{R_1} \text{ if } |E_{out}| \leq V_Z + 0.7\ V$$

where V_Z = Zener breakdown voltage

Table 12-2 (cont.)

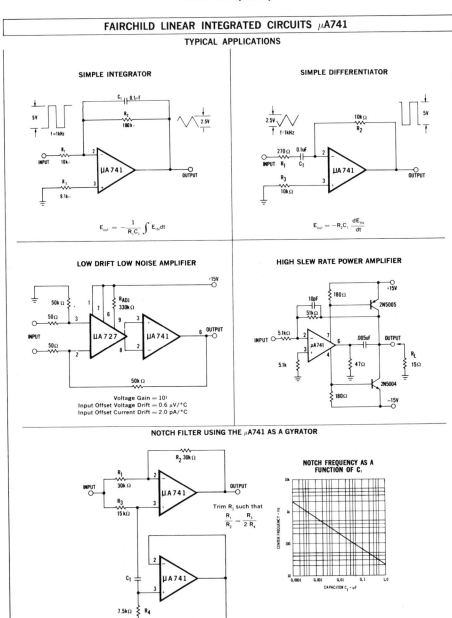

(Courtesy of Fairchild Semiconductor.)

The 30 pF capacitor between the bases of $Q14$ and $Q16$ is the basis of the internal frequency compensation. The inclusion of this high voltage dielectric, high-valued (relatively speaking) capacitor in a monolithic circuit is the result of new improved processing techniques. Its inclusion internal to the operational amplifier package results in increased reliability, more compact circuit boards, and lower assembly costs. If an application does require reduced high-frequency response (for reasons other than instability), the connection of capacitance between the output and $Q6$'s emitter (pins 5 and 6) will suffice.

The electrical characteristics of the $\mu A741$ are presented on page 292. Notice that they are provided in duplicate, once at room temperature, 25°C, and again for over the entire device operating range of -55 to $+125°C$. This device has a very high open-loop gain (A_o) of 200,000, typically with a minimum value of 50,000.

Following the electrical characteristics are a rather complete set of performance curves on pages 292–294. The student will learn a great deal about operational amplifiers if each of these curves is analyzed and understood. With this information the $\mu A741$ can be successfully applied to virtually any application. The specification sheet for the $\mu A741$ is completed on pages 295–296 with some definitions and applications. Notice the additional resistor equal to the parallel combination of $R1$ and $R2$ in the inverting and non-inverting amplifier circuits. It is added to minimize output errors due to input offset currents and is required in high-accuracy applications.

PROBLEMS

1. Determine an expression for e_{out} for the circuit shown in Fig. P12-1.

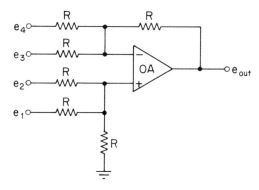

FIG. P12-1.

2. Figure P12-2 shows a current-to-voltage operational amplifier circuit. Develop an expression for e_{out} as a function of i_{in}.

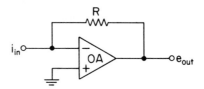

FIG. P12-2. Current-to-voltage circuit

3. Develop an operational amplifier circuit to function as a Wien bridge oscillator.

4. The waveform shown in Fig. P12-4 is applied to the inverting input of a comparator circuit. The noninverting input is grounded and ± 12 V power supplies are used. Sketch e_{out} as a function of time.

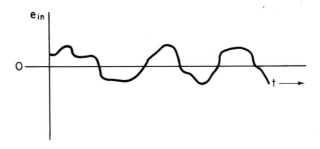

FIG. P12-4.

5. Figure P12-5 shows a simple multiplier circuit. The two photoconductive cells are equally coupled to the light output of lamp $L1$. Describe how an output proportional to the product of two other signals is obtained.

6. Determine an expression for e_o for the circuit shown in Fig. P12-6. Why is the circuit known as a scaling adder?

7. What is the function of the 910 Ω resistor in Fig. P12-6 and how does it accomplish that function?

8. Design an operational amplifier circuit that will function as a gyrator.

9. An operational amplifier has a bandwidth of 1 MHz and a slew rate of 2 V/μs. Determine the maximum sinusoidal output at 1 MHz.

10. What is the highest frequency at which a 10 V p-p sinusoidal output is possible with the operational amplifier described in Problem 9?

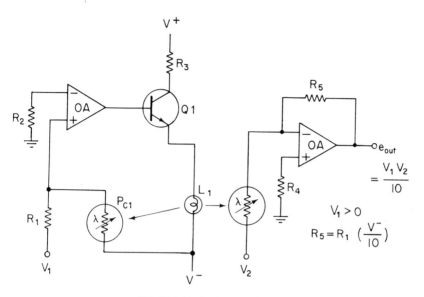

FIG. P12-5. Analog multiplier

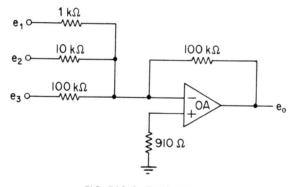

FIG. P12-6. Scaling adder

13

Survey of Communication Systems

13-1 Introduction

The ultimate goal of the circuits we have studied in this text is to apply them in usable fashion. The major applications of the circuits are in communication systems. These systems had their beginning with the discovery of various electrical, magnetic, and electrostatic phenomena prior to the twentieth centry. Lee De Forest's investigations of the triode vacuum tube in the early 1900s allowed for the first form of electronic amplification and opened the door to wireless communication. In 1948 another major discovery in the history of electronics occurred with the development of the transistor by Shockley, Brittain, and Bardeen.

The function of a communication system is to transfer information from one point to another via some communications link. The very first form of "information" transferred was the human voice or a code, which was then converted back to words. Man had a natural desire to communicate rapidly between any points on the earth and that initially was the major concern of these developments. As that goal became a reality with the evolution of new technology following the invention of the triode vacuum tube, new and less basic applications were also realized, such as entertainment, radar, and radio telescopes. The communications field is still a highly dynamic one with new semiconductor devices constantly making new equipment possible or allowing improvement of the old units. Communications was the basic origin of electronics, and no other major field developed until the transistor made modern digital computers a reality. We now have two major subcategories in the field of electronics—communications and digital systems.

Basic to the field of communications is the concept of modulation. *Modulation* is the process of impressing information onto a high-frequency carrier for transmission. It follows then that once this information is received the intelligence must be removed from the high-frequency carrier—a process known as *demodulation*. At this point you may be thinking why bother to go through this modulation–demodulation process—why not just transmit the information directly and save the bother? The problem is that the frequency of the human voice ranges from about 20–4000 Hz. If everyone transmitted those frequencies directly as radio waves, interference between them would cause them all to be ineffective. As it turns out, it is virtually impossible to transmit such low frequencies anyway, since the required antennas for efficient propagation would have to be miles in length.

The answer to these problems then is modulation, which allows long-distance propagation of the low-frequency intelligence with a high-frequency carrier. The high frequency carriers are chosen such that only one transmitter in an area operates at the same frequency to eliminate interference, and that frequency is high enough such that antenna sizes are physically small and manageable. There are three possible methods of impressing low frequency information onto a higher frequency. Equation (13-1) is the mathematical representation of a sine wave, which we shall assume to be our high frequency carrier:

$$v = V_P \sin(\omega t + \Phi) \qquad (13\text{-}1)$$

where v = instantaneous value
V_P = peak value
ω = angular velocity = $2\pi f$
Φ = phase angle

Any one of the last three terms could be varied in accordance with the low frequency information signal so as to produce a *modulated signal*. If the amplitude term, V_P, is the parameter varied, it is termed *amplitude modulation* (AM). If the frequency is varied, it is termed *frequency modulation* (FM). Varying the phase angle Φ results in *phase modulation* (PM).

Communication systems are often categorized by the frequency of the carrier. Table 13-1 provides the names for the various ranges of frequencies in the radio spectrum. Note the acronym abbreviation in the last column.

The extra high frequency range begins at the starting point of infrared

Table 13-1 Radio-Frequency Spectrum

FREQUENCY	DESIGNATION	ABBREVIATION
3–30 kHz	Very low frequency	VLF
30–300 kHz	Low frequency	LF
300 kHz–3 MHz	Medium frequency	MF
3–30 MHz	High frequency	HF
30–300 MHz	Very high frequency	VHF
300 MHz–3 GHz	Ultra high frequency	UHF
3–30 GHz	Super high frequency	SHF
30–300 GHz	Extra high frequency	EHF

frequencies, but the infrareds extend considerably beyond 300 GHz. After the infrareds in the electromagnetic spectrum (of which the radio waves are a very small portion) come light waves, ultraviolet, X-rays, gamma rays, and cosmic rays. None of these frequencies (above 300 GHz) are classified as radio frequencies.

13-2 Communication Systems and Noise Effects

A communication system can be very simple, but it can also assume very complex proportions. Figure 13-1 represents a simple system in block-diagram form. Notice that the *modulator* accepts two inputs, the *carrier* and the *information* signal. It produces the *modulated signal*, which is subsequently amplified before transmission. Transmission can take place by any one of three means—*antennas, waveguides,* or *transmission lines.* The receiving unit of the system then picks up the transmitted signal, but must reamplify it to compensate for attenuation that occurred during its transmission. Once suitably amplified it is fed to the demodulator (often referred to as the detector) where the information signal is extracted from the high-frequency carrier and then fed to the power amplifier. The signal is brought to a suitably high level by the power amplifier to drive a speaker or any other load (output transducer).

A very important aspect of the field of electronic communications is the study of electrical noise. It may be defined as any form of energy tending to interfere with the reception and reproduction of electronic communication signals. It can be broadly categorized into two subgroups: that which is introduced in the transmitting medium (usually the atmosphere), and internal noise created within the transmitter and receiver themselves.

The *external noise* affecting the reception of radio signals is caused by

304 *Survey of Communication System*

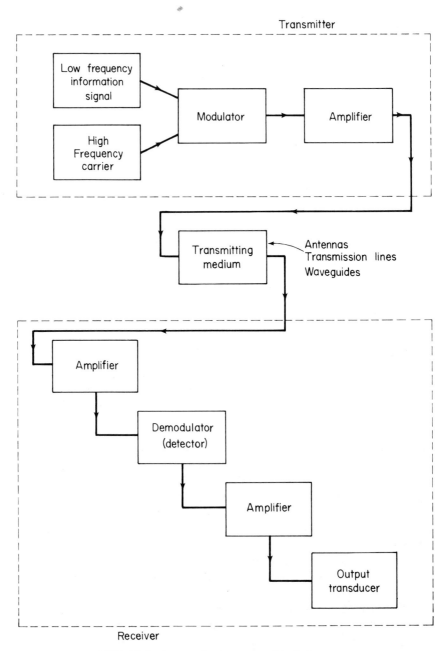

FIG. 13-1. Communication system block diagram

several different factors. *Atmospheric noise* is caused by lightning discharges and other naturally occurring electrical disturbances (such as cosmic radiation) in the atmosphere. It is generally heard as *static* in a radio receiver. Its frequency content is spread rather uniformly throughout the frequency spectrum up to about 30 MHz, above which it begins tapering off. Another source of external noise is the sun, and its emissions are termed solar noise. The sun is a variable source of noise, which reaches a peak of activity every 11 years. It affects some forms of communications drastically during these periods. The next peak in its emissions is due in 1979. Other stars besides our sun emit such noise and this is referred to as cosmic noise. Space noise is not significant below a frequency of 20 MHz or above 1.43 GHz (1 GHz = 10^9 Hz).

One final form of external noise is man-made noise. It is significant between the frequencies of 1 and 600 MHz in most industrial areas. It is the dominant noise factor (considering both external and internal effects) in that range and is caused by automotive and other ignition systems, switching of reactive loads, and leakage from high voltage power transmission lines. It is obviously a variable noise source that is difficult to analyze other than by statistical means.

Internal noise is created by both active and passive electronic devices. It is of a truly random character and hence is analyzed statistically. This noise is spread evenly in frequency throughout the radio spectrum. There are two major sources of internal noise, shot noise and thermal agitation noise. Thermal agitation noise is also referred to as *white* or *Johnson* noise. It is generated in the resistance of any device, active or passive, and is caused by the rapid random motion of its molecules, atoms, and electrons. The root-mean-square value of the voltage this noise produces is predicted by

$$E_N = 4kT\delta f R \qquad (13\text{-}2)$$

where $k =$ Boltzman's constant $= 1.38 \times 10^{-23}$ joule (J)/°K
 $T =$ absolute temperature, °K $= 273° +$ °C
 $\delta f =$ bandwidth of interest
 $R =$ value of resistance

We thus see that thermal agitation noise increases with temperature, bandwidth of the system, and resistance value. The peak value of this noise cannot be accurately predicted but seldom exceeds 10 times the calculated root-mean-square value. This noise, although usually small, can be bothersome at the front end of a high gain receiver.

Shot noise is the random variation in the arrival of electrons (or holes) at the collector of a transistor. It also occurs in virtually all other electronic active devices. The paths taken by electrons (or holes) are random, thus leading to unwanted variations in current flow. Since it is such a difficult noise to analyze, the device manufacturers will often provide a resistance value that generates an equivalent level of noise as the device's shot noise.

The ability of a communication system to minimize the effects of electrical noise to a great extent determines the receiver's sensitivity. The *sensitivity* may be defined as the minimum level of signal at the receiver's input that will produce an acceptable output signal at the speaker or headphones. This is not to be confused with a receiver's *selectivity*, which is its ability to discriminate between desired and undesirable frequencies.

13-3 Amplitude Modulation

Amplitude modulation (AM) is the process of varying the amplitude of a high frequency sine wave (the carrier) in relation to the intelligence (modulation) signal. The amplitude of the carrier is made proportional to the instantaneous value of the modulating voltage. It is produced by passing the carrier and modulating signal through a nonlinear device. A transistor operating with very low bias or high bias has a nonlinear characteristic as shown in Fig. 13-2. The result of passing two sine waves through a nonlinear device is shown in Fig. 13-3, and the output signal is an AM wave. Notice its amplitude is varying at a frequency equal to the frequency of the modulating signal input. The AM waveform is made up of three sinusoidal components, one of them at the carrier frequency and two sidebands. The upper sideband is at the carrier plus the modulating frequency, and the lower sideband is at the carrier minus the modulating frequency.

An AM waveform can mathematically be described as

$$e = E_c \sin \omega_c t + \frac{mE_c}{2} \cos(\omega_c - \omega_m)t - \frac{mE_c}{2} \cos(\omega_c + \omega_m)t \quad (13\text{-}3)$$

where E_c = peak value of carrier
ω_c = carrier's angular velocity
ω_m = modulating signal's angular velocity
m = the modulation index

The *modulation index*, m, is always ≤ 1 for AM. It is also expressed as a percentage and is then known as the percentage of modulation. It expresses a ratio of modulation signal level to carrier signal, and at 100 per cent modulation (maximum) the E_{min} level shown in Fig. 13-3 is zero. The value of m can be calculated from the max and min values shown in Fig. 13-3 as follows:

$$m = \frac{E_{max} - E_{min}}{E_{max} + E_{min}} \quad (13\text{-}4)$$

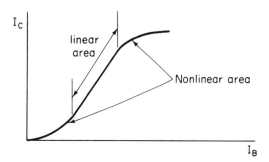

FIG. 13-2. Nonlinear transistor characteristics

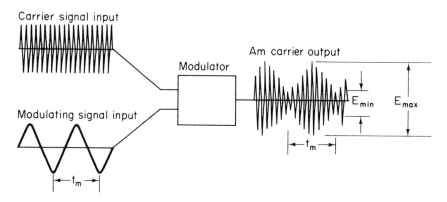

FIG. 13-3. Generation of AM waveform

It is often of interest to be able to calculate the amount of power contained in the sidebands as compared to the carrier. The following relationship allows for that calculation:

$$\frac{P_T}{P_C} = 1 + \frac{m^2}{2} \tag{13-5}$$

where P_T = total transmitted and P_C = carrier power. Once this ratio is calculated, the power in the carrier is

$$P_C = P_T - P_{SB} \tag{13-6}$$

and the power in the upper sideband equals the power in the lower sideband and they are equal. Hence

$$P_{USB} = P_{LSB} = \frac{P_{SB}}{2} \tag{13-7}$$

308 Survey of Communication Systems

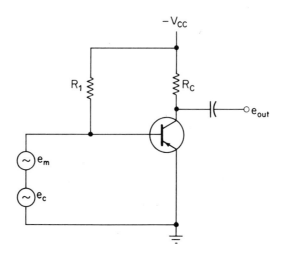

FIG. 13-4. Simple AM modulator

Amplitude-modulated waveforms are usually generated by operating transistors in a nonlinear area. Figure 13-4 shows a very simple and yet effective means of accomplishing this purpose. By making $R1$ extremely large, the transistor is at very low bias (close to cutoff) and thus nonlinear. Other possibilities for AM generation include injecting one of the two signals at the base and the other at the emitter, and vice versa. Since many AM transmitters must generate high power outputs, they must use tubes in at least the output stages, since solid-state technology has not yet advanced very far into the high-power high-frequency field.

Once the AM signal is generated, transmitted, received, and amplified, it becomes necessary to somehow extract the intelligence information from the signal. The circuit used almost exclusively to perform this function is the diode detector circuit shown in Fig. 13-5. The diode rectifies the incoming

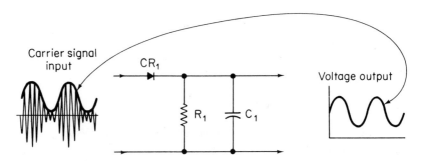

FIG. 13-5. Amplitude-modulated detector circuit

AM signal, leaving only the positive portion of the waveform, while the *RC* circuit effectively filters out the waveform's high frequency content. This then leaves only the outline of the positive portion of the AM wave, which is exactly the original modulating signal, as is desired. This detected signal is then suitably amplified to drive the final output transducer.

13-4 Frequency Modulation

Since phase modulation (PM) is so similar to frequency modulation (FM) and also since PM involves theory difficult to comprehend, we shall study FM only in this survey of communications. In an FM system it is the frequency of the carrier that is made to vary in accordance with the modulating signal. This seems simple enough, but several subtleties cause the student to often confuse the issue. The *rate* at which the FM carrier signal is made to vary is the rate (frequency) of the modulating signal, and the *amount* of deviation from the original carrier frequency is made proportional to the strength (amplitude) of the modulating signal. Now carefully reread the previous sentence and then consider the simple FM transmitter shown in Fig. 13-6.

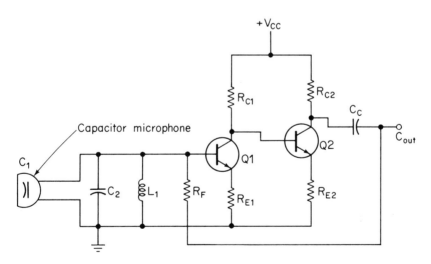

FIG. 13-6. Simple FM transmitter

The transmitter shown in Fig. 13-6 utilizes a capacitor microphone to generate an FM signal and is very helpful in aiding the student to understand the concept of FM generation. Let us say that you whistle into the microphone at a frequency of 1000 Hz. The sound waves striking the surface plate of the microphone cause its effective capacitance to vary at the frequency of the modulating signal, your whistle. Since the $Q1-Q2$ amplifier combination forms a simple Franklin-type oscillator circuit, it is seen that the frequency of oscillation is being changed at a 1000-Hz rate and will be rising and falling around the rest (carrier) frequency, since the whistle will cause the microphone's capacitance to rise and fall around its rest capacitance.

The loudness of your whistle will also have an effect on this circuit. The loudness will affect the *amount* of capacitance change and therefore determine the amount of deviation from the center, rest, or carrier frequency, depending on how you wish to refer to it. Figure 13-7 provides a visualization of the

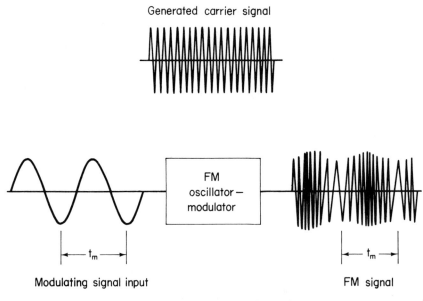

FIG. 13-7. Frequency-modulated signal

FM-generated waveform. Notice that the FM signal's amplitude is nearly constant (ideally it is perfectly constant). Its frequency, however, is varying at the frequency of the modulating signal.

Other, more complex, methods are usually used to generate FM than the capacitor microphone method just explained because of greater efficiency (greater frequency shift for a given amplitude-modulating signal). Regardless of the method of generation, the signal is ultimately amplified, transmitted, received, reamplified, and then must be detected. The FM slope detector is the most simple method available, and it involves converting the FM signal into an AM signal and then using the simple AM diode detector explained in Section 13-3. The conversion from FM to AM is accomplished by feeding the FM signal through a tank circuit tuned to a frequency somewhat above (or below) the FM carrier frequency. The tank output signal will then vary in amplitude according to the FM signal's frequency. This occurs because the tank presents a variable impedance to the FM signal, as shown in Fig. 13-8, and thus the output varies with frequency. This allows the original sinusoidal modulating signal to be reproduced at the detector's output. Figure 13-9 provides a circuit that performs this function with the waveforms shown to help you visualize the role of the three functions performed by the slope detector circuit.

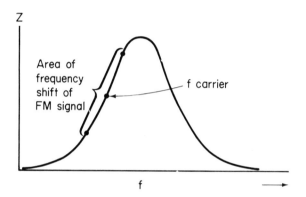

FIG. 13-8. Z versus f for a parallel tank circuit

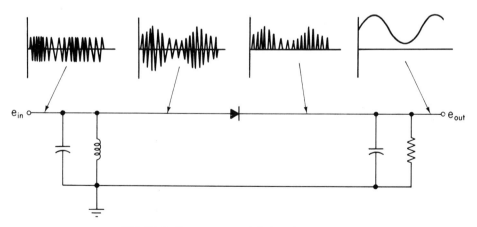

FIG. 13-9. Frequency-modulation slope detector

Transmission of radio signals via FM techniques offers an advantage over AM systems with respect to minimization of noise effects. Since the receiver usually has a circuit to limit the signal to a constant amplitude, the effect of spurious noise pulses above that level is minimized. A more subtle source of noise reduction in FM receivers also occurs, but is beyond the scope of this text. It is, however, even more effective than the limiter action in reducing noise effects.

The fidelity (quality) of reproduction on the standard FM stations (88–108 MHz) is better than standard AM stations (540–1600 kHz) mainly because of the greater bandwidth allocated for the FM stations. The FM stations operate within a 200 kHz bandwidth; AM stations use only 10 kHz. Thus FM offers no great advantage over AM from an inherent standpoint with respect to fidelity.

13-5 Transmission Methods—Antennas

The *wavelength* of any given frequency may be defined as the distance its wave travels in the time it takes to complete one cycle. Since radio waves travel at nearly the speed of light, c (3×10^8 meters (m)/s), the wavelength of any radio wave can be found by the following mathematical relationship:

$$\text{wavelength (in meters)} = \lambda = \frac{c}{f} \qquad (13\text{-}8)$$

For an antenna to effectively radiate or receive a radio wave, it is usually necessary for it to be at least $\lambda/4$ in length or greater. An exception to this rule is the ferrite core loopstick antenna common to most standard AM radio receivers. Since the standard AM broadcast band is such a low frequency, by radio standards, its wavelengths are quite long. For instance, at 1000 kHz we can calculate $\lambda/4$ as

$$\frac{\lambda}{4} = \frac{c}{4f} = \frac{3 \times 10^8 \text{ m/s}}{4 \times 1 \times 10^6 \text{ Hz}} = 75 \text{ m}$$

This means that the transmitting antenna for this station would be around 75 m tall (over 200 ft), which of course is out of the question for the receiver.

The antenna's function is to take the electrical energy from the transmitter and convert it into electromagnetic energy, which is capable of traveling through the atmosphere and most other dielectrics. The receiving antenna's function is to reverse that process—to reconvert the electromagnetic energy into electrical energy. Generally, transmitting and receiving antennas are so similar that a *principle of reciprocity* exists. It states that all characteristics of antennas are identical whether they are being used for transmission or reception.

If an alternating current is passed through a conductor, a magnetic field (H) is set up around the wire and an electric field (E) is set up from one end to the other, as shown in Fig. 13-10. These two mutually perpendicular

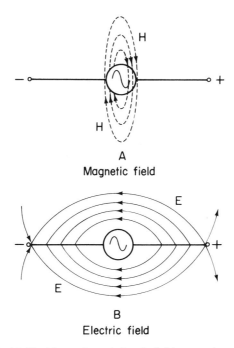

FIG. 13-10. Magnetic and electric fields around a wire

fields alternately receive energy from the electrical ac source as they build up, and return energy to the electrical source as they collapse. Whenever the electrical length of the wire is an appreciable fraction of the frequency's wavelength, a strange phenomenon takes place. The electromagnetic field does not collapse fully, but instead becomes propagated into space. The reciprocity principle then tells us the reverse of this process occurs if a similar antenna is placed in the path of this radio wave. This receiving antenna then has a voltage and current induced in it, which is then suitably amplified by the receiver circuitry.

The usual method by which these electromagnetic waves get from point

A to point B is any one of the following:

1. Ground (surface) waves.
2. Sky-wave propagation—the ionosphere.
3. Space waves.

The waves from an antenna travel into space in all directions. Those which travel along the ground's surface are generally affected by the terrain features, and are called *ground waves*. These waves are attenuated by the earth's surface in varying degrees, depending upon the earth's conductivity and the frequency of the wave. Thus the effective usable distance of transmission depends on the transmitted power level, the terrain, the frequency, and of course the receiver's sensitivity. This mode of propagation is normally useful at very low frequencies due to the otherwise high attenuation.

Sky waves depend upon the refractive effects on the transmitted wave by the various gases in the earth's upper atmosphere (the ionosphere). They serve to bend some radio waves around back toward the earth. The various layers of the ionosphere are dependent upon radiation from the sun, and hence this form of propagation varies between day and night, and with variations in the sun's radiation. They are labeled the D, E, F_1, F_2, and F layers, as shown in Fig. 13-11. The figure also shows the elimination of the D and E

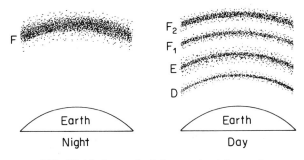

FIG. 13-11. Ionospheric layers about the earth

layers and the splitting of the F layer into the F_1 and F_2 layers at night. Their refractive abilities are reduced as the transmitter's frequency is increased, until a point is reached where the wave passes on through the ionosphere into free space and is no longer sent back to earth.

Figure 13-12 shows that lower frequencies radiated from a transmitter are refracted back to earth before higher ones. Reflections from the earth back to the ionosphere are shown in Fig. 13-13 and are known as *multiple-hop transmission*. The figure also shows the same signal "launched" from the antenna at different angles (as usually happens) and arriving at the same receiving site. If they arrive out of phase because of the different distances traveled, distortion will result in the receiver's output.

Space waves behave very simply compared to sky waves. They travel directly (in relatively straight lines) between the transmitting and receiving antennas. They are therefore limited by the curvature of the earth and the height of the antennas. The following equation approximates this effective distance as

$$d \simeq \sqrt{2H_t} + \sqrt{2H_R} \tag{13-9}$$

where d = useful distance of propagation in miles
H_t = height of transmitting antenna in feet
H_R = height of receiving antenna in feet

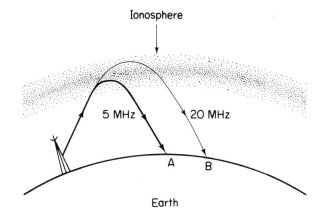

FIG. 13-12. Ionospheric refraction differences with frequency

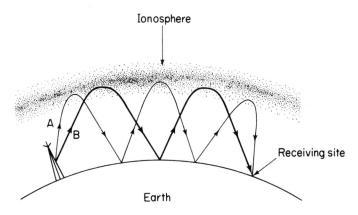

FIG. 13-13. Multiple-hop transmission

Space-wave propagation is the usual mode for VHF frequencies and above, since they are too high to be refracted by the ionosphere and they are too greatly attenuated as ground waves. This is the case for standard FM and TV broadcasts and explains the approximate 100 mile limit for their reception.

As you know, antennas come in a wide variety of shapes and sizes. This is to provide the best characteristics for a given application. An analysis of the many varieties of transmitting and receiving antennas is beyond this book's intentions.

13-6 Transmission Methods—Transmission Lines and Waveguides

Although antennas are the most common method of transmitting radio energy, transmission lines and waveguides are also often used. The mode of energy transmission chosen for a given application would normally depend on the following factors:

1. Initial cost and long-term maintenance.
2. Frequency band and information—carrying capacity.
3. Selectivity or privacy offered.
4. Reliability and noise characteristics.
5. Power level and efficiency.

Naturally, any one mode of energy transmission will have only some of the desirable features. It therefore becomes a matter of sound technical judgement to choose the mode of energy transmission best suited for a particular application. The following examples show that each method of transmission has its proper place.

It is desired to transmit a 100 MHz signal between two points 30 miles apart. If received energy in each case is chosen as 1 nanowatt (nW), the required transmitted power for the frequency given is

1. Transmission lines: 10^{1500} megawatts (MW) (15,000 dB loss).
2. Waveguides: 10^{150} MW (1500 dB loss).
3. Antennas: 100 milliwatts (mW) (80 dB loss).

Clearly, the transmission of energy without any electrical conductors (antennas) will be found to exceed the efficiency of waveguides and transmission lines by many orders of magnitude. As a practical consequence of these results, microwave antenna relay links at about 30 mile intervals are used for cross-country transmission of telephone and television services.

On the other hand, if the transmission path length of the preceding example is reduced by a factor of 0.01 to a distance of 1500 ft, the comparison yields

1. Transmission line: 1 MW (150 dB loss).
2. Waveguide: 30 nW (15 dB loss).
3. Antenna: 10 microwatts (μW) (40 dB loss).

Clearly, over short distances the waveguide now excels over both transmission line and antenna for efficiency of energy transfer.

A comparison of the energy limit required to obtain a received power of 1 nW for the three modes of energy transmission is shown in Fig. 13-14. The

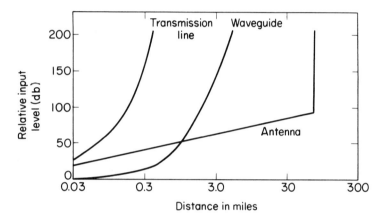

FIG. 13-14. Input power required versus distance for fixed receiver power and transmission frequency

frequency is 1000 MHz, and the results are expressed on a decibel scale with a 0 dB reference at the required receiver power level of 1 nW. The dotted section of the antenna curve, somewhat beyond 30 miles, indicates that the attenuation becomes severe beyond the line-of-sight distance, which is typically 50 miles.

One final comparison, and then we shall look at waveguides and transmission lines. Transmission of energy down to zero frequency is practical with transmission lines, but waveguides and antennas inherently have a practical low-frequency limit. In the case of antennas, this limit is about 100 kHz, and for waveguides it is about 300 MHz. Theoretically antennas and waveguides could be made to work at arbitrarily low frequencies, but the physical sizes required would become excessively large. However, with the low gravity and lack of atmosphere on the moon, it may be feasible to have an antenna 10 miles high and 100 miles long for frequencies as low as a few hundred hertz. As an indication of the sizes involved, it may be noted that for either waveguides or antennas the important dimension is normally one half-wavelength. Thus a waveguide for a 300 MHz signal would be about the size of a roadway drainage culvert, and an antenna for 300 MHz would be about $1\frac{1}{2}$ ft long.

A transmission line is simply two electrical conductors used to transmit electrical energy. Thus any two conductors can be technically considered to be a transmission line. However, the consideration that separates transmission line study from simple conductors is the effects of operation at high frequencies. At high frequencies some new line characteristics are encountered other than just the resistive losses considered at lower frequencies.

An important concept in the study of transmission lines is the characteristic impedance of the line. The *characteristic impedance*, Z_o, is defined as the impedance seen looking into an infinitely long line. This impedance can be calculated based upon the physical dimensions of the line. Figure 13-15 shows

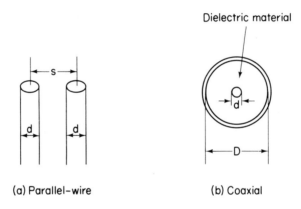

(a) Parallel-wire   (b) Coaxial

FIG. 13-15. Transmission-line types and geometry

the parallel-wire and coaxial types of transmission line and their important dimensions. The characteristic impedance of the parallel-wire transmission line shown in Fig. 13-15a is

$$Z_o = 276 \log \frac{2s}{d} \; \Omega \qquad (13\text{-}10)$$

where s and d are as noted in Fig. 13-15a.

The coaxial line shown in Fig. 13-15b consists of an inner conductor supported by some dielectric material and an outer shield conductor that is usually grounded. Its characteristic impedance is

$$Z_o = \frac{138}{K} \log \frac{D}{d} \; \Omega \qquad (13\text{-}11)$$

where $K =$ the dielectric constant of the dielectric and D and d are as noted in Fig. 13-15b.

A transmission line may be simulated at high frequencies by the LC circuit shown in Fig. 13-16. This is a somewhat idealized approach since no resistive losses are shown, but this is usually a justifiable simplification. The

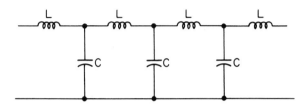

FIG. 13-16. Transmission-line RF equivalent circuit

values of L and C for a given line are usually specified per unit length, because they occur continuously along the line. They cannot be assumed to be lumped at any one point. It can be mathematically shown that a line's characteristic impedance is given by

$$Z_o = \sqrt{\frac{L}{C}}\,\Omega \qquad (13\text{-}12)$$

and thus still another means of determining Z_o is available to us.

If a line is terminated in a resistance equal to the line's characteristic impedance, all the energy transmitted by a generator will be absorbed at the load. In addition, the line's input impedance will still be Z_o in spite of the fact that the line is not infinite in length. Unfortunately, the load to which a line is delivering RF energy cannot often be made exactly equal to the purely resistive characteristic impedance. Under those conditions a portion of the energy delivered to the load is reflected back towards the generator. Methods of minimizing this inefficiency are the basic goal of transmission line work.

At low frequencies we are only concerned with the variations of voltage with time along a conductor. At high frequencies, however, the voltage along a conductor (transmission line) varies with distance also, since the line length is usually an appreciable fraction of a wavelength. Recall that a sine wave goes through one complete cycle in a distance of one wavelength. *Therefore, in transmission line work we must be concerned with voltage variations with time at one specific point and also with voltage variations with location at one specific instant of time.*

A waveguide may be crudely thought of as a coaxial transmission line with the center conductor removed. The energy is propagated in the electromagnetic mode as with antennas, but instead of radiating into space the wave is contained by the metallic inner walls of the waveguide. Waveguides are seldom used below 3000 MHz because of size considerations. The most common shape is rectangular, and the longer of the two inner dimensions must be at least a half-wavelength. Circular waveguides are also commonly used. At high frequencies waveguides have less attenuation than transmission lines and can handle much higher power levels.

Index

A

α, 8
Adder, 280, 281
AM (amplitude modulation), 306-9
Amplification, 13-17
Amplifiers:
  audio, 172
  basic, 63-85
  class A, 152
  class AB, 152
  class B, 152, 158-64
  dc, 172
  differential, 259-60, 273-74
  distortion, 147-51
  FET, 113-38
  large-signal, 141
  logarithmic, 284
  multistage, 201-21
  operational, 273-97
  paraphase, 162
  power (*see* Power amplifier)
  small signal, 141
  tuned, 190-98
  video, 172
Amplitude distortion, 148
Amplitude modulation (AM), 306-9
Antenna, 312-16
Atmospheric noise, 305
Audio amplifiers, 172, 216-21
Avalanche diode (*see* Zener diode)
Average current:
  diode, 25
  full-wave, 36
  half-wave, 25

B

β, 9
Bandpass filter, 176
Bandwidth, 193
Barkhausen criterion, 228

Barrier potential, 3
Bipolar devices, 113
Breakdown diode (*see* Zener diode)
Breakdown voltage, 4
Bridge oscillator (*see* Wein-bridge oscillator)
Bridge rectifier, 28-30
Butterworth arrangement, 198
Bypass capacitor, 100-103

C

Capacitance multiplier, 257-58
CB (common-base) amplifier, 63-72
CC (common-collector) amplifier, 72-79
CE (common-emitter) amplifier, 79-85, 87-109
Center-tap rectifier, 28-31
Channel, 114
Characteristic impedance, 318-19
Clapp oscillator, 232-33
Class A, 152-58
Class AB, 164-69
Class B, 158-64
Closed-loop gain, 89
Collector, 7
Colpitts oscillator, 230-33
Common-base (CB) amplifier, 63-72
Common-collector (CC) amplifier (*see* Emitter follower)
Common-emitter (CE) amplifier, 79-85, 87-109
Common-mode effects, 289-90
Comparator, 279-80
Complementary pair, 163
Complementary symmetry, 163
Constant current supply, 57-60
Continuous wave, 228
Crossover distortion, 163-64
Crystal, quartz, 233-34
Crystal oscillator, 233-35
Current gain, 14
Current limiting, 53-57
Current source, 57-60

D

Darlington compound, 202-6
dc level shifter, 255-57
Decibels, 173-75
Decoupling, 244
Demodulation, 308-9
Depletion mode, 120
Depletion region, 115
Derating, 146
Detector, 308, 309
Differential amplifier, 259-60, 273-74
Differentiator, 283
Diffusion, 252
Diode, 2-6
Diode curve, 4
Diode detector, 308-9
Diode resistance, 5
Distortion:
  amplitude, 148
  crossover, 163-64
  FET, 137-38
  frequency, 148
  with negative feedback, 151
  phase, 148
Distortion analysis, 148-51
Doping, 2
Doubler, voltage, 38-39
Drain resistance, 116
Dynamic drain resistance, 116

E

Emitter bypass capacitor, 100-103
Emitter feedback biasing, 97-100
Emitter follower, 72-79, 152
Enhancement mode, 120
External noise, 303-5

F

Feedback:
  amplifier, 87-109

Feedback (cont.)
  collector, 103-7
  control system, 50
  distortion, 151
  emitter, 91-94
  local, 209
  multistage, 209-12
  negative, 88, 130-32
  oscillators, 228-33
  positive, 228
  principles, 88-91
  voltage divider, 107-9
FET (field-effect transistor):
  biasing, 122-32
  definitions, 119
  distortion, 137-38
  J, 113-19
  MOS, 113, 120-22
  power applications, 138
  square law relationship, 136-37
Filter:
  bandpass, 176
  factor, 88
  high-pass, 172-73
  low-pass, 175
  power-supply, 31-37
FM (frequency modulation), 309-12
Follower circuit, 279
Forward bias, 3
Franklin oscillator, 229-30
Frequency distortion, 148
Frequency modulation, 309-12
Frequency response (*see* Bandwidth; High-frequency analysis; Low-frequency analysis)
Full-wave rectifier, 28-31

G

Gain impedance relations, 17
Gain variations, 91-94
Ground wave, 314
Gyrator, 260-62

H

h_{fb}, 8
h_{ib}, 67
h_{fe}, 9-13
Half-wave rectifier, 24-27
Harmonic distortion, 148-51
Hartley oscillator, 230-33
Heat dissipation, 143-47
Heat sink, 144
High frequency analysis:
 bipolar, 183-90
 FET, 180-83
Hole, definition, 2
Hybrid integrated circuits, 247-51

I

I_{DSS}, 118-19
Idle current, 165
IF (intermediate-frequency) amplifier, 195-96
IGFET (insulated-gate FET), 113
Inductance simulation, 260-62
Input offset current, 289
Input offset voltage, 289
Integrated circuits (IC), 247-70
Integrator, 282-83
Internal noise, 305
Inverting input, 274

J

JFET (junction FET), 113-19
Johnson noise, 305
Junction field-effect transistor (JFET), 113-19
Junction resistance, 5

L

Large-scale integration (LSI), 248-50
Large-signal amplifier, 141 (*see also* Harmonic distortion)

Line regulation, 42
Linear amplifier, 198
Linear circuit, 1
Loading effect, 204
Load regulation, 41
Logarithmic amplifier, 284
Low-frequency analysis, 176-80
Low-pass filter, 172-76
LSI (large-scale integration), 248-50

M

Medium-scale integration (MSI), 248-50
Metal-oxide semiconductor FET (MOSFET), 113, 120-22
Midband, 171
Miller effect, 181-84
Modulation, 302, 306-12
Modulation index, 306
Monolithic (integrated circuits), 247, 251-53 (*see also* Integrated circuits)
MOSFET (metal-oxide semiconductor FET), 113, 120-22
Multiple-hop transmission, 314-15
Multiplier:
   capacitor, 257, 258
   voltage, 38-39
Multistage amplifiers, 201-21

N

Negative feedback (*see* Feedback)
Negative resistance oscillators, 240-43
Noise, 303-6
Noninverting input, 274
npn transistor, 7

O

ωCR product, 34-35
Offset current, 289
Offset voltage, 289
Ohmic region, 115
Open-loop gain, 88

Operating point, 8, 92-93
Operational amplifier (op amp), 273-97
Oscillations, unwanted, 243-44
Oscillators:
   crystal, 233-35
   LC, 226, 228-33
   RC, 226, 235-40
   types of:
      Clapp, 232-33
      Colpitts, 230-33
      crystal, 233-35
      Franklin, 229-30
      Hartley, 230-33
      negative resistance, 240-43
      Pierce, 234-35
      Wien-bridge, 237-40
Output impedance:
   CC, 74
   CE, 80
   source follower, 133-34
Overcurrent protection, 53-56

P

Paraphase amplifier, 162
Parasitic oscillations, 243-44
Phase distortion, 148
Phase inverter (*see* Phase splitter)
Phase modulation, 309
Phase shift oscillator, 235-37
Phase splitter, 162
Pierce crystal oscillator, 234-35
Pinch-off, 113
pn junction, 2-6
pnp transistor, 7
Positive feedback, 228
Power amplifier:
   definition, 141
   design example, 216-21
   efficiency, 153-55
   integrated circuit, 267-70
Power gain:
   definition, 14

Power supply, 23-57 (*see also* Rectifier; Ripple; Voltage regulator)
Power-supply filter, 31-37
Power-supply ripple reduction, 51-52
p-type semiconductor, 2
Pulsating current, 31
Push-pull amplifier, 158-68

Q

Q, 191-94
Q point drift, JFET, 124
Quality factor, 191
Quartz crystal, 233-34
Quiescent (Q) point, 8, 92-93

R

Radio frequency amplifier, 172
RC filters, 172-76
Reciprocity, 313
Rectifier:
  bridge, 28-30
  center-tap, 28-29
  full-wave, 28-30
  half-wave, 24-27
  voltage doubler, 38-39
  (*see also* Power-supply filter)
Rectifier diode, 2-6, 23-30 (*see also* Zener diode)
Regenerative feedback (*see* Feedback, oscillators; Positive feedback)
Regulated power supply, 41-42, 46-67
Regulation, 42
Reverse bias, 4
Reverse resistance of diode, 5
Ripple:
  definition, 31
  factor, 34
  reduction with regulated supply, 51-52

S

S-factor, 96
Self-bias with JFET, 125
Series regulator, 49-57

Shockley's relation, 67
Shot noise, 305-6
Shunt regulator, 46-49
Sky wave, 314
Slew rate, 288-89
Slope detector, 311
Source follower, 132-36
Space-wave, 315-16
Stability:
 bias, 97-100
 factor, 96
Stagger tuning, 196-98
Static, 305
Subtractor, 281-82
Summing amplifier, 280-81

T

Tank circuit, 91
TGIR, 18
Thermal noise, 305
Thermal resistance, 144
Thick-film circuit, 247-51
Thin-film circuit, 247-51
Transadmittance, 117
Transconductance, 116-17
Transformer-coupled Class A amplifier, 156-58
Transistor:
 derating, 146-47
Transistor gain-impedance relationship, 17-19
Transmission line, 316-20
Tuned amplifier, 190-98
Tuned circuit, 191
Tunnel diode, 240-43

U

Unipolar device, definition, 113

V

Video amplifier, 172
Voltage-divider feedback, 107-9

Voltage gain, definition, 14
Voltage multipliers, 38-39
Voltage regulator, 41-57
Voltage source, 42-46

W

Waveguide, 319-20
Wavelength, 312
White noise, 305
Wien-bridge oscillator, 237-40

Z

Zener diode, 46-49
Zener impedance, 46
Zener regulator, 46-57
Zero bias drain current, IDSS, 118-19